Tests and Worksheets

SAXON Math™
HOMESCHOOL
6/5

Stephen Hake
John Saxon

SAXON™
PUBLISHERS

Saxon Publishers gratefully acknowledges the contributions of the following individuals in the completion of this project:

Authors: Stephen Hake, John Saxon

Editorial: Chris Braun, Bo Björn Johnson, Brooke Butner, Matt Maloney, Brian E. Rice

Editorial Support Services: Christopher Davey, Jay Allman, Shelley Turner, Jean Van Vleck, Darlene Terry

Production: Alicia Britt, Karen Hammond, Donna Jarrel, Brenda Lopez, Adriana Maxwell, Cristi D. Whiddon

Project Management: Angela Johnson, Becky Cavnar

Printed in the United States of America

ISBN: 978-1-59-141322-6

34   0928   21

4500826119

C O N T E N T S

Introduction

Saxon Math 6/5—Homeschool Tests and Worksheets contains Facts Practice Tests, Activity Sheets, tests, and recording forms. Brief descriptions of these components are provided below, and additional information can be found on the pages that introduce each section. Solutions to the Facts Practice Tests, Activity Sheets, and tests are located in the *Saxon Math 6/5—Homeschool Solutions Manual.* For a complete overview of the philosophy and implementation of Saxon Math™, please refer to the preface of the *Saxon Math 6/5—Homeschool* textbook.

About the Facts Practice Tests

Facts Practice Tests are an essential and integral part of Saxon Math™. Mastery of basic facts frees your student to focus on procedures and concepts rather than computation. Employing memory to recall frequently encountered facts permits students to bring higher-level thinking skills to bear when solving problems.

Facts Practice Tests should be administered as prescribed at the beginning of each lesson or test. Sufficient copies of the Facts Practice Tests for one student are supplied, in the order needed, with the corresponding lesson or test clearly indicated at the top of the page. Limit student work on these tests to five minutes or less. Your student should keep track of his or her times and scores and get progressively faster and more accurate as the course continues.

About the Activity Sheets

Selected lessons and investigations in the student textbook present content through activities. These activities often require the use of worksheets called Activity Sheets, which are provided in this workbook in the quantities needed by one student.

About the Tests

The tests are designed to reward your student and to provide you with diagnostic information. Every lesson in the student textbook culminates with a cumulative mixed practice, so the tests are cumulative as well. By allowing your student to display his or her skills, the tests build confidence and motivation for continued learning. The cumulative nature of Saxon tests also gives your student an incentive to master skills and concepts that might otherwise be learned for just one test.

All the tests needed for one student are provided in this workbook. The testing schedule is printed on the page immediately preceding the first test. Administering the tests according to the schedule is essential. Each test is written to follow a specific five-lesson interval in the textbook. Following the schedule allows your student to gain sufficient practice on new topics before being tested over them.

About the Recording Forms

The last section of this book contains five optional recording forms. Three of the forms provide an organized framework for your student to record his or her work on the daily lessons, Mixed Practices, and tests. Two of the forms help track and analyze your student's performance on his or her assignments. All five of the recording forms may be photocopied as needed.

Facts Practice Tests
and Activity Sheets

This section contains the Facts Practice Tests and Activity Sheets, which are sequenced in the order of their use in *Saxon Math 6/5—Homeschool*. Sufficient copies for one student are provided.

Facts Practice Tests

Rapid and accurate recall of basic facts and skills dramatically increases students' mathematical abilities. To that end we have provided the Facts Practice Tests. Begin each lesson with the Facts Practice Test suggested in the Warm-Up, limiting the time to five minutes or less. Your student should work independently and rapidly during the Facts Practice Tests, trying to improve on previous performances in both speed and accuracy.

Each Facts Practice Test contains a line for your student to record his or her time. Timing the student is motivating. Striving to improve speed helps students automate skills and offers the additional benefit of an up-tempo atmosphere to start the lesson. Time invested in practicing basic facts is repaid in your student's ability to work faster.

After each Facts Practice Test, quickly read aloud the answers from the *Saxon Math 6/5—Homeschool Solutions Manual* as your student checks his or her work. If your student made any errors or was unable to finish within the allotted time, he or she should correct the errors or complete the problems as part of the day's assignment. You might wish to have your student track Facts Practice scores and times on Recording Form A, which is found in this workbook.

On test day the student should be held accountable for mastering the content of recent Facts Practice Tests. Hence, each test identifies a Facts Practice Test to be taken on that day. Allow five minutes on test days for the student to complete the Facts Practice Test before beginning the cumulative test.

Activity Sheets

Activity Sheets are referenced in certain lessons and investigations of *Saxon Math 6/5—Homeschool*. Students should refer to the textbook for detailed instructions on using the Activity Sheets. The fraction manipulatives (on Activity Sheets 13–18) may be color-coded with colored pencils or markers before they are cut out.

A — 100 Addition Facts
For use with Lesson 1

Name _____

Time _____

(handwritten: 60 probs / 5.5 min.)

Add.

3 + 2 **5**	8 + 3 **11**	2 + 1 **3**	5 + 6 **11**	2 + 9 **11**	4 + 8 **12**	8 + 0 **8**	3 + 9 **12**	1 + 0 **1**	6 + 3 **9**
7 + 3 **10**	1 + 6 **7**	4 + 7	0 + 3 **3**	6 + 4	5 + 5 **10**	3 + 1 **4**	7 + 2	8 + 5	2 + 5 **7**
4 + 0 **9**	5 + 7	1 + 1 **2**	5 + 4 **9**	2 + 8 **10**	7 + 1 **8**	4 + 6 **10**	0 + 2 **2**	6 + 5	4 + 9
8 + 6	0 + 4 **9**	5 + 8	7 + 4	1 + 7 **8**	6 + 6 **19**	4 + 1 **5**	8 + 2 **10**	2 + 4 **6**	6 + 0 **6**
9 + 1 **10**	8 + 8	2 + 2 **4**	4 + 5 **9**	6 + 2	0 + 0 **0**	5 + 9	3 + 3 **6**	8 + 1 **4**	2 + 7
4 + 4 **8**	7 + 5	0 + 1 **1**	8 + 7	3 + 4	7 + 9	1 + 2 **3**	6 + 7	0 + 8 **8**	9 + 2 **11**
0 + 9 **9**	8 + 9	7 + 6	1 + 3	6 + 8	2 + 0 **2**	8 + 4	3 + 5	9 + 8	5 + 0 **5**
9 + 3 **12**	2 + 6	3 + 0 **3**	6 + 1 **7**	3 + 6	5 + 2	0 + 5 **5**	6 + 9	1 + 8 **0**	9 + 6 **15**
4 + 3	9 + 9 **18**	0 + 7 **7**	9 + 4 **13**	7 + 7	1 + 4 **5**	3 + 7	7 + 0 **7**	2 + 3	5 + 1 **6**
9 + 5 **8**	1 + 5 **9**	9 + 0	3 + 8	1 + 9 **10**	5 + 3	4 + 2	9 + 7 **16**	0 + 6 **6**	7 + 8

A   100 Addition Facts

For use with Lesson 2

Name _____

Time _____

Add.

3 + 2	8 + 3	2 + 1	5 + 6	2 + 9	4 + 8	8 + 0	3 + 9	1 + 0	6 + 3
7 + 3	1 + 6	4 + 7	0 + 3	6 + 4	5 + 5	3 + 1	7 + 2	8 + 5	2 + 5
4 + 0	5 + 7	1 + 1	5 + 4	2 + 8	7 + 1	4 + 6	0 + 2	6 + 5	4 + 9
8 + 6	0 + 4	5 + 8	7 + 4	1 + 7	6 + 6	4 + 1	8 + 2	2 + 4	6 + 0
9 + 1	8 + 8	2 + 2	4 + 5	6 + 2	0 + 0	5 + 9	3 + 3	8 + 1	2 + 7
4 + 4	7 + 5	0 + 1	8 + 7	3 + 4	7 + 9	1 + 2	6 + 7	0 + 8	9 + 2
0 + 9	8 + 9	7 + 6	1 + 3	6 + 8	2 + 0	8 + 4	3 + 5	9 + 8	5 + 0
9 + 3	2 + 6	3 + 0	6 + 1	3 + 6	5 + 2	0 + 5	6 + 9	1 + 8	9 + 6
4 + 3	9 + 9	0 + 7	9 + 4	7 + 7	1 + 4	3 + 7	7 + 0	2 + 3	5 + 1
9 + 5	1 + 5	9 + 0	3 + 8	1 + 9	5 + 3	4 + 2	9 + 7	0 + 6	7 + 8

Saxon Math 6/5—Homeschool

100 Addition Facts
For use with Lesson 3

Name _____

Time _____

Add.

3 + 2	8 + 3	2 + 1	5 + 6	2 + 9	4 + 8	8 + 0	3 + 9	1 + 0	6 + 3
7 + 3	1 + 6	4 + 7	0 + 3	6 + 4	5 + 5	3 + 1	7 + 2	8 + 5	2 + 5
4 + 0	5 + 7	1 + 1	5 + 4	2 + 8	7 + 1	4 + 6	0 + 2	6 + 5	4 + 9
8 + 6	0 + 4	5 + 8	7 + 4	1 + 7	6 + 6	4 + 1	8 + 2	2 + 4	6 + 0
9 + 1	8 + 8	2 + 2	4 + 5	6 + 2	0 + 0	5 + 9	3 + 3	8 + 1	2 + 7
4 + 4	7 + 5	0 + 1	8 + 7	3 + 4	7 + 9	1 + 2	6 + 7	0 + 8	9 + 2
0 + 9	8 + 9	7 + 6	1 + 3	6 + 8	2 + 0	8 + 4	3 + 5	9 + 8	5 + 0
9 + 3	2 + 6	3 + 0	6 + 1	3 + 6	5 + 2	0 + 5	6 + 9	1 + 8	9 + 6
4 + 3	9 + 9	0 + 7	9 + 4	7 + 7	1 + 4	3 + 7	7 + 0	2 + 3	5 + 1
9 + 5	1 + 5	9 + 0	3 + 8	1 + 9	5 + 3	4 + 2	9 + 7	0 + 6	7 + 8

1

One-Dollar Bills

For use with Lesson 3

2

One-Dollar Bills

For use with Lesson 3

3   Ten-Dollar Bills

For use with Lesson 3

4

Ten-Dollar Bills
For use with Lesson 3

5

One Hundred–Dollar Bills

For use with Lesson 3

6 One Hundred–Dollar Bills

For use with Lesson 3

7 | Place-Value Template
For use with Lesson 3

PLACE-VALUE TEMPLATE

ones

tens

hundreds

A — 100 Addition Facts
For use with Lesson 4

Name _____

Time _____

Add.

3 + 2 = 5	8 + 3 = 11	2 + 1 = 3	5 + 6 = 11	2 + 9 = 11	4 + 8 = 12	8 + 0 = 8	3 + 9 = 12	1 + 0 = 1	6 + 3 = 9
7 + 3 = 10	1 + 6 = 7	4 + 7 = 11	0 + 3 = 3	6 + 4 = 10	5 + 5 = 10	3 + 1 = 4	7 + 2 = 9	8 + 5 = 13	2 + 5 = 7
4 + 0 = 4	5 + 7 = 12	1 + 1 = 2	5 + 4 = 9	2 + 8 = 10	7 + 1 = 8	4 + 6 = 10	0 + 2 = 2	6 + 5 = 11	4 + 9 = 13
8 + 6 = 14	0 + 4 = 4	5 + 8 = 13	7 + 4 = 11	1 + 7 = 8	6 + 6 = 15	4 + 1 = 5	8 + 2 = 10	2 + 4 = 6	6 + 0 = 6
9 + 1 = 10	8 + 8 = 16	2 + 2 = 4	4 + 5 = 6	6 + 2 = 8	0 + 0 = 0	5 + 9 = 14	3 + 3 = 2	8 + 1 = 9	2 + 7 = 9
4 + 4 = 8	7 + 5 = 15	0 + 1 = 1	8 + 7 = 15	3 + 4 = 7	7 + 9 = 12	1 + 2 = 3	6 + 7 = 12	0 + 8 = 8	9 + 2 = 11
0 + 9 = 9	8 + 9 = 17	7 + 6 = 13	1 + 3 = 4	6 + 8 = 14	2 + 0 = 2	8 + 4 = 12	3 + 5 = 8	9 + 8 = 17	5 + 0 = 5
9 + 3 = 12	2 + 6 = 8	3 + 0 = 3	6 + 1 = 7	3 + 6 = 9	5 + 2 = 7	0 + 5 = 5	6 + 9 = 15	1 + 8 = 9	9 + 6 = 15
4 + 3 = 7	9 + 9 = 18	0 + 7 = 7	9 + 4 = 13	7 + 7 = 14	1 + 4 = 5	3 + 7 = 10	7 + 0 = 7	2 + 3 = 5	5 + 1 = 6
9 + 5 = 14	1 + 5 = 6	9 + 0 = 9	3 + 8 = 11	1 + 9 = 10	5 + 3 = 8	4 + 2 = 2	9 + 7 = 12	0 + 6 = 6	7 + 8 =

A 100 Addition Facts

For use with Lesson 5

Name _____

Time _____

Add.

3 + 2	8 + 3	2 + 1	5 + 6	2 + 9	4 + 8	8 + 0	3 + 9	1 + 0	6 + 3
7 + 3	1 + 6	4 + 7	0 + 3	6 + 4	5 + 5	3 + 1	7 + 2	8 + 5	2 + 5
4 + 0	5 + 7	1 + 1	5 + 4	2 + 8	7 + 1	4 + 6	0 + 2	6 + 5	4 + 9
8 + 6	0 + 4	5 + 8	7 + 4	1 + 7	6 + 6	4 + 1	8 + 2	2 + 4	6 + 0
9 + 1	8 + 8	2 + 2	4 + 5	6 + 2	0 + 0	5 + 9	3 + 3	8 + 1	2 + 7
4 + 4	7 + 5	0 + 1	8 + 7	3 + 4	7 + 9	1 + 2	6 + 7	0 + 8	9 + 2
0 + 9	8 + 9	7 + 6	1 + 3	6 + 8	2 + 0	8 + 4	3 + 5	9 + 8	5 + 0
9 + 3	2 + 6	3 + 0	6 + 1	3 + 6	5 + 2	0 + 5	6 + 9	1 + 8	9 + 6
4 + 3	9 + 9	0 + 7	9 + 4	7 + 7	1 + 4	3 + 7	7 + 0	2 + 3	5 + 1
9 + 5	1 + 5	9 + 0	3 + 8	1 + 9	5 + 3	4 + 2	9 + 7	0 + 6	7 + 8

A | **100 Addition Facts**
For use with Lesson 6

Name _____

Time _____

Add.

3 + 2	8 + 3	2 + 1	5 + 6	2 + 9	4 + 8	8 + 0	3 + 9	1 + 0	6 + 3
7 + 3	1 + 6	4 + 7	0 + 3	6 + 4	5 + 5	3 + 1	7 + 2	8 + 5	2 + 5
4 + 0	5 + 7	1 + 1	5 + 4	2 + 8	7 + 1	4 + 6	0 + 2	6 + 5	4 + 9
8 + 6	0 + 4	5 + 8	7 + 4	1 + 7	6 + 6	4 + 1	8 + 2	2 + 4	6 + 0
9 + 1	8 + 8	2 + 2	4 + 5	6 + 2	0 + 0	5 + 9	3 + 3	8 + 1	2 + 7
4 + 4	7 + 5	0 + 1	8 + 7	3 + 4	7 + 9	1 + 2	6 + 7	0 + 8	9 + 2
0 + 9	8 + 9	7 + 6	1 + 3	6 + 8	2 + 0	8 + 4	3 + 5	9 + 8	5 + 0
9 + 3	2 + 6	3 + 0	6 + 1	3 + 6	5 + 2	0 + 5	6 + 9	1 + 8	9 + 6
4 + 3	9 + 9	0 + 7	9 + 4	7 + 7	1 + 4	3 + 7	7 + 0	2 + 3	5 + 1
9 + 5	1 + 5	9 + 0	3 + 8	1 + 9	5 + 3	4 + 2	9 + 7	0 + 6	7 + 8

A

100 Addition Facts
For use with Lesson 7

Name _____

Time ∞

Add.

3 + 2 **5**	8 + 3 **11**	2 + 1 **3**	5 + 6 **11**	2 + 9 **11**	4 + 8 **15**	8 + 0 **8**	3 + 9 **15**	1 + 0 **1**	6 + 3 **9**
7 + 3 **10**	1 + 6 **7**	4 + 7 **11**	0 + 3 **3**	6 + 4 **10**	5 + 5 **10**	3 + 1 **4**	7 + 2 **9**	8 + 5 **13**	2 + 5 **7**
4 + 0 **4**	5 + 7 **12**	1 + 1 **2**	5 + 4 **9**	2 + 8 **10**	7 + 1 **8**	4 + 6 **10**	0 + 2 **2**	6 + 5 **11**	4 + 9 **13**
8 + 6 **14**	0 + 4 **4**	5 + 8 **15**	7 + 4 **11**	1 + 7 **8**	6 + 6 **10**	4 + 1 **5**	8 + 2 **10**	2 + 4 **6**	6 + 0 **6**
9 + 1 **10**	8 + 8 **16**	2 + 2 **4**	4 + 5 **9**	6 + 2 **8**	0 + 0 **0**	5 + 9 **14**	3 + 3 **6**	8 + 1 **8**	2 + 7 **9**
4 + 4 **8**	7 + 5 **15**	0 + 1 **1**	8 + 7 **15**	3 + 4 **7**	7 + 9 **16**	1 + 2 **8**	6 + 7 **15**	0 + 8 **8**	9 + 2 **11**
0 + 9 **9**	8 + 9 **17**	7 + 6 **13**	1 + 3 **4**	6 + 8 **14**	2 + 0 **2**	8 + 4 **15**	3 + 5 **8**	9 + 8 **17**	5 + 0 **5**
9 + 3 **12**	2 + 6 **8**	3 + 0 **3**	6 + 1 **7**	3 + 6 **9**	5 + 2 **7**	0 + 5 **5**	6 + 9 **15**	1 + 8 **9**	9 + 6 **15**
4 + 3 **7**	9 + 9 **18**	0 + 7 **7**	9 + 4 **13**	7 + 7 **14**	1 + 4 **5**	3 + 7 **10**	7 + 0 **2**	2 + 3 **15**	5 + 1 **6**
9 + 5 **14**	1 + 5 **2**	9 + 0 **9**	3 + 8 **11**	1 + 9 **10**	5 + 3 **8**	4 + 2 **16**	9 + 7 **6**	0 + 6 **6**	7 + 8 **15**

Saxon Math 6/5—Homeschool

FACTS PRACTICE TEST

100 Addition Facts
For use with Lesson 8

Name _____

Time _____

Add.

3 + 2	8 + 3	2 + 1	5 + 6	2 + 9	4 + 8	8 + 0	3 + 9	1 + 0	6 + 3
7 + 3	1 + 6	4 + 7	0 + 3	6 + 4	5 + 5	3 + 1	7 + 2	8 + 5	2 + 5
4 + 0	5 + 7	1 + 1	5 + 4	2 + 8	7 + 1	4 + 6	0 + 2	6 + 5	4 + 9
8 + 6	0 + 4	5 + 8	7 + 4	1 + 7	6 + 6	4 + 1	8 + 2	2 + 4	6 + 0
9 + 1	8 + 8	2 + 2	4 + 5	6 + 2	0 + 0	5 + 9	3 + 3	8 + 1	2 + 7
4 + 4	7 + 5	0 + 1	8 + 7	3 + 4	7 + 9	1 + 2	6 + 7	0 + 8	9 + 2
0 + 9	8 + 9	7 + 6	1 + 3	6 + 8	2 + 0	8 + 4	3 + 5	9 + 8	5 + 0
9 + 3	2 + 6	3 + 0	6 + 1	3 + 6	5 + 2	0 + 5	6 + 9	1 + 8	9 + 6
4 + 3	9 + 9	0 + 7	9 + 4	7 + 7	1 + 4	3 + 7	7 + 0	2 + 3	5 + 1
9 + 5	1 + 5	9 + 0	3 + 8	1 + 9	5 + 3	4 + 2	9 + 7	0 + 6	7 + 8

FACTS PRACTICE TEST

For use with Lesson 9

B 100 Subtraction Facts

Name _____

Time _____

Subtract.

$16 - 9 = 7$	$7 - 1 = 2$	$18 - 9 = 9$	$11 - 3$	$13 - 7$	$8 - 2$	$11 - 5 = 8$	$5 - 0 = 5$	$17 - 9$	$6 - 1$
$10 - 9$	$6 - 2$	$13 - 4$	$4 - 0 = 4$	$10 - 5 = 5$	$5 - 1 = 4$	$10 - 3 = 7$	$12 - 6 = 6$	$10 - 1 = 9$	$6 - 4$
$7 - 2$	$14 - 7 = 7$	$8 - 1 = 7$	$11 - 6$	$3 - 3 = 0$	$16 - 7$	$5 - 2 = 3$	$12 - 4$	$3 - 0 = 3$	$11 - 7$
$17 - 8 = 6$	$6 - 0 = 4$	$10 - 6 = 3$	$4 - 1$	$9 - 5$	$9 - 0 = 9$	$5 - 4$	$12 - 5$	$4 - 2$	$9 - 3 = 5$
$12 - 3 = 9$	$16 - 8 = 8$	$9 - 1 = 8$	$15 - 6$	$11 - 4$	$13 - 5$	$1 - 0 = 1$	$8 - 5$	$9 - 6$	$11 - 2 = 9$
$7 - 0 = 7$	$10 - 8 = 5$	$6 - 3$	$14 - 5$	$3 - 1 = 2$	$8 - 6$	$4 - 4 = 0$	$11 - 8 = 3$	$3 - 2 = 1$	$15 - 9$
$13 - 8$	$7 - 4$	$10 - 7 = 3$	$0 - 0 = 0$	$12 - 8$	$5 - 5 = 0$	$4 - 3 = 1$	$8 - 7$	$7 - 3$	$7 - 6 = 1$
$5 - 3 = 5$	$7 - 5 = 5$	$2 - 1 = 1$	$6 - 6 = 0$	$8 - 4 = 4$	$2 - 2 = 0$	$13 - 6$	$15 - 8$	$2 - 0 = 2$	$13 - 9$
$1 - 1 = 0$	$11 - 9 = 2$	$10 - 4 = 0$	$9 - 2$	$14 - 6$	$8 - 0 = 8$	$9 - 4$	$10 - 2 = 8$	$6 - 5$	$8 - 3$
$7 - 7 = 0$	$14 - 8$	$12 - 9$	$9 - 8 = 1$	$12 - 7$	$9 - 9 = 0$	$15 - 7$	$8 - 8 = 0$	$14 - 9$	$9 - 7$

Saxon Math 6/5 — Homeschool

B — 100 Subtraction Facts

For use with Lesson 10

Name _____

Time _____

Subtract.

16 − 9 = **7**	7 − 1 = **2**	18 − 9 = **9**	11 − 3 = **8**	13 − 7 = **6**	8 − 2 =	11 − 5 =	5 − 0 = **5**	17 − 9 =	6 − 1 = **5**
10 − 9 = **1**	6 − 2 =	13 − 4 =	4 − 0 = **4**	10 − 5 = **5**	5 − 1 = **4**	10 − 3 = **7**	12 − 6 =	10 − 1 = **10**	6 − 4 =
7 − 2 = **8**	14 − 7 =	8 − 1 = **7**	11 − 6 =	3 − 3 = **0**	16 − 7 =	5 − 2 =	12 − 4 =	3 − 0 = **3**	11 − 7 =
17 − 8 =	6 − 0 = **6**	10 − 6 = **8**	4 − 1 = **3**	9 − 5 = **4**	9 − 0 = **9**	5 − 4 =	12 − 5 = **7**	4 − 2 = **2**	9 − 3 =
12 − 3 = **9**	16 − 8 = **8**	9 − 1 = **8**	15 − 6 =	11 − 4 = **7**	13 − 5 =	1 − 0 = **1**	8 − 5 =	9 − 6 =	11 − 2 = **9**
7 − 0 = **7**	10 − 8 = **2**	6 − 3 =	14 − 5 = **9**	3 − 1 = **2**	8 − 6 = **5**	4 − 4 = **0**	11 − 8 = **3**	3 − 2 =	15 − 9 =
13 − 8 =	7 − 4 =	10 − 7 = **3**	0 − 0 = **0**	12 − 8 =	5 − 5 = **0**	4 − 3 = **1**	8 − 7 =	7 − 3 =	7 − 6 =
5 − 3 = **5**	7 − 5 =	2 − 1 = **1**	6 − 6 = **0**	8 − 4 =	2 − 2 = **0**	13 − 6 = **7**	15 − 8 =	2 − 0 = **2**	13 − 9 =
1 − 1 = **0**	11 − 9 =	10 − 4 = **2**	9 − 2 = **7**	14 − 6 =	8 − 0 = **8**	9 − 4 =	10 − 2 = **8**	6 − 5 =	8 − 3 =
7 − 7 = **0**	14 − 8 =	12 − 9 = **1**	9 − 8 =	12 − 7 =	9 − 9 = **0**	15 − 7 = **8**	8 − 8 = **0**	14 − 9 = **5**	9 − 7 = **5**

A   100 Addition Facts
For use with Test 1

Name _____

Time _____

Add.

3 + 2	8 + 3	2 + 1	5 + 6	2 + 9	4 + 8	8 + 0	3 + 9	1 + 0	6 + 3
7 + 3	1 + 6	4 + 7	0 + 3	6 + 4	5 + 5	3 + 1	7 + 2	8 + 5	2 + 5
4 + 0	5 + 7	1 + 1	5 + 4	2 + 8	7 + 1	4 + 6	0 + 2	6 + 5	4 + 9
8 + 6	0 + 4	5 + 8	7 + 4	1 + 7	6 + 6	4 + 1	8 + 2	2 + 4	6 + 0
9 + 1	8 + 8	2 + 2	4 + 5	6 + 2	0 + 0	5 + 9	3 + 3	8 + 1	2 + 7
4 + 4	7 + 5	0 + 1	8 + 7	3 + 4	7 + 9	1 + 2	6 + 7	0 + 8	9 + 2
0 + 9	8 + 9	7 + 6	1 + 3	6 + 8	2 + 0	8 + 4	3 + 5	9 + 8	5 + 0
9 + 3	2 + 6	3 + 0	6 + 1	3 + 6	5 + 2	0 + 5	6 + 9	1 + 8	9 + 6
4 + 3	9 + 9	0 + 7	9 + 4	7 + 7	1 + 4	3 + 7	7 + 0	2 + 3	5 + 1
9 + 5	1 + 5	9 + 0	3 + 8	1 + 9	5 + 3	4 + 2	9 + 7	0 + 6	7 + 8

Stories About Combining   Name _____

For use with Investigation 1

(a) The troop hiked 8 miles in the morning.

(b) The troop hiked 7 miles in the afternoon.

(c) Altogether, the troop hiked 15 miles in the morning and afternoon.

Complete the question that replaces the missing statement in each of the following paragraphs.

1. Sentence (a) is missing:
    The troop hiked 7 miles in the afternoon. Altogether, the troop hiked 15 miles in the morning and afternoon.

    How many miles _____

    _____

    _____

2. Sentence (b) is missing:
    The troop hiked 8 miles in the morning. Altogether, the troop hiked 15 miles in the morning and afternoon.

    How many miles _____

    _____

    _____

3. Sentence (c) is missing:
    The troop hiked 8 miles in the morning. The troop hiked 7 miles in the afternoon.

    Altogether, how many miles _____

    _____

    _____

9 **Stories About Separating** Name _____

For use with Investigation 1

(d) Jack went to the store with $28.

(e) Jack spent $12 at the store.

(f) Jack left the store with $16.

Complete the question that replaces the missing statement in each of the following paragraphs.

4. Sentence (d) is missing:

Jack spent $12 at the store. Jack left the store with $16.

How much money _____

_____

_____

5. Sentence (e) is missing:

Jack went to the store with $28. Jack left the store with $16.

How much money _____

_____

_____

6. Sentence (f) is missing:

Jack went to the store with $28. Jack spent $12 at the store.

How much money _____

_____

_____

10 Stories About Equal Groups

For use with Investigation 1

Name _____

(g) At the Lazy W Ranch there are 4 pens.

(h) There are 30 cows in each pen.

(i) Altogether, there are 120 cows at the Lazy W Ranch.

Complete the question that replaces the missing statement in each of the following paragraphs.

7. Sentence (g) is missing:

There are 30 cows in each pen. Altogether, there are 120 cows at the Lazy W Ranch.

How many pens _____

_____

_____

8. Sentence (h) is missing:

At the Lazy W Ranch there are 4 pens. Altogether, there are 120 cows at the Lazy W Ranch.

How many cows _____

_____

_____

9. Sentence (i) is missing:

At the Lazy W Ranch there are 4 pens. There are 30 cows in each pen.

Altogether, how many _____

_____

_____

11 | Stories About Comparing Name _____

For use with Investigation 1

(j) Abe is 5 years old.

(k) Gabe is 11 years old.

(l) Version 1: Gabe is 6 years older than Abe.

(l) Version 2: Abe is 6 years younger than Gabe.

Complete the question that replaces the missing statement in each of the following paragraphs.

10. Sentence (j) is missing:

Gabe is 11 years old. Gabe is 6 years older than Abe.

How old _____

_____

11. Sentence (k) is missing:

Abe is 5 years old. Abe is 6 years younger than Gabe.

How old _____

_____

12. Sentence (l), version 1 is missing:

Abe is 5 years old. Gabe is 11 years old.

Gabe is _____

_____

13. Sentence (l), version 2 is missing:

Abe is 5 years old. Gabe is 11 years old.

Abe is _____

_____

FACTS PRACTICE TEST

B | 100 Subtraction Facts
For use with Lesson 11

Name _____

Time _____

Subtract.

16 − 9	7 − 1 *6*	18 − 9 *9*	11 − 3	13 − 7	8 − 2 *6*	11 − 5 *6*	5 − 0 *5*	17 − 9	6 − 1 *5*
10 − 9 *1*	6 − 2 *4*	13 − 4	4 − 0 *4*	10 − 5 *5*	5 − 1 *4*	10 − 3	12 − 6	10 − 1 *9*	6 − 4 *2*
7 − 2	14 − 7 *7*	8 − 1 *7*	11 − 6	3 − 3 *0*	16 − 7	5 − 2 *3*	12 − 4	3 − 0 *3*	11 − 7
17 − 8	6 − 0 *6*	10 − 6 *9*	4 − 1 *3*	9 − 5 *4*	9 − 0 *9*	5 − 4	12 − 5	4 − 2 *2*	9 − 3
12 − 3	16 − 8	9 − 1 *8*	15 − 6	11 − 4	13 − 5	1 − 0 *1*	8 − 5	9 − 6	11 − 2
7 − 0 *7*	10 − 8 *2*	6 − 3 *3*	14 − 5	3 − 1 *2*	8 − 6	4 − 4	11 − 8	3 − 2 *1*	15 − 9
13 − 8	7 − 4	10 − 7	0 − 0 *0*	12 − 8	5 − 5 *0*	4 − 3	8 − 7	7 − 3	7 − 6
5 − 3 *2*	7 − 5	2 − 1	6 − 6 *0*	8 − 4	2 − 2 *0*	13 − 6	15 − 8	2 − 0 *2*	13 − 9
1 − 1 *0*	11 − 9	10 − 4 *6*	9 − 2	14 − 6	8 − 0 *8*	9 − 4 *5*	10 − 2 *8*	6 − 5 *1*	8 − 3
7 − 7 *0*	14 − 8	12 − 9	9 − 8	12 − 7	9 − 9 *0*	15 − 7	8 − 8 *0*	14 − 9	9 − 7

| B |

100 Subtraction Facts
For use with Lesson 12

Name _____

Time _____

Subtract.

16 − 9	7 − 1	18 − 9	11 − 3	13 − 7	8 − 2	11 − 5	5 − 0	17 − 9	6 − 1
10 − 9	6 − 2	13 − 4	4 − 0	10 − 5	5 − 1	10 − 3	12 − 6	10 − 1	6 − 4
7 − 2	14 − 7	8 − 1	11 − 6	3 − 3	16 − 7	5 − 2	12 − 4	3 − 0	11 − 7
17 − 8	6 − 0	10 − 6	4 − 1	9 − 5	9 − 0	5 − 4	12 − 5	4 − 2	9 − 3
12 − 3	16 − 8	9 − 1	15 − 6	11 − 4	13 − 5	1 − 0	8 − 5	9 − 6	11 − 2
7 − 0	10 − 8	6 − 3	14 − 5	3 − 1	8 − 6	4 − 4	11 − 8	3 − 2	15 − 9
13 − 8	7 − 4	10 − 7	0 − 0	12 − 8	5 − 5	4 − 3	8 − 7	7 − 3	7 − 6
5 − 3	7 − 5	2 − 1	6 − 6	8 − 4	2 − 2	13 − 6	15 − 8	2 − 0	13 − 9
1 − 1	11 − 9	10 − 4	9 − 2	14 − 6	8 − 0	9 − 4	10 − 2	6 − 5	8 − 3
7 − 7	14 − 8	12 − 9	9 − 8	12 − 7	9 − 9	15 − 7	8 − 8	14 − 9	9 − 7

Saxon Math 6/5—Homeschool

B **100 Subtraction Facts**
For use with Lesson 13

Name _____

Time _____

Subtract.

16 − 9	7 − 1	18 − 9	11 − 3	13 − 7	8 − 2	11 − 5	5 − 0	17 − 9	6 − 1
10 − 9	6 − 2	13 − 4	4 − 0	10 − 5	5 − 1	10 − 3	12 − 6	10 − 1	6 − 4
7 − 2	14 − 7	8 − 1	11 − 6	3 − 3	16 − 7	5 − 2	12 − 4	3 − 0	11 − 7
17 − 8	6 − 0	10 − 6	4 − 1	9 − 5	9 − 0	5 − 4	12 − 5	4 − 2	9 − 3
12 − 3	16 − 8	9 − 1	15 − 6	11 − 4	13 − 5	1 − 0	8 − 5	9 − 6	11 − 2
7 − 0	10 − 8	6 − 3	14 − 5	3 − 1	8 − 6	4 − 4	11 − 8	3 − 2	15 − 9
13 − 8	7 − 4	10 − 7	0 − 0	12 − 8	5 − 5	4 − 3	8 − 7	7 − 3	7 − 6
5 − 3	7 − 5	2 − 1	6 − 6	8 − 4	2 − 2	13 − 6	15 − 8	2 − 0	13 − 9
1 − 1	11 − 9	10 − 4	9 − 2	14 − 6	8 − 0	9 − 4	10 − 2	6 − 5	8 − 3
7 − 7	14 − 8	12 − 9	9 − 8	12 − 7	9 − 9	15 − 7	8 − 8	14 − 9	9 − 7

B   100 Subtraction Facts
For use with Lesson 14

Name _____

Time _____

Subtract.

16 − 9	7 − 1 *6*	18 − 9	11 − 3	13 − 7	8 − 2	11 − 5	5 − 0	17 − 9	6 − 1 *5*
10 − 9	6 − 2	13 − 4	4 − 0 *4*	10 − 5 *5*	5 − 1 *4*	10 − 3	12 − 6	10 − 1 *9*	6 − 4
7 − 2	14 − 7	8 − 1 *7*	11 − 6	3 − 3 *0*	16 − 7	5 − 2	12 − 4	3 − 0 *3*	11 − 7
17 − 8	6 − 0 *6*	10 − 6	4 − 1 *3*	9 − 5 *4*	9 − 0 *9*	5 − 4 *1*	12 − 5	4 − 2	9 − 3
12 − 3	16 − 8	9 − 1 *8*	15 − 6	11 − 4	13 − 5	1 − 0 *1*	8 − 5	9 − 6 *3*	11 − 2
7 − 0 *7*	10 − 8	6 − 3	14 − 5	3 − 1 *2*	8 − 6	4 − 4 *0*	11 − 8	3 − 2 *1*	15 − 9
13 − 8	7 − 4	10 − 7	0 − 0 *0*	12 − 8	5 − 5 *0*	4 − 3 *1*	8 − 7 *1*	7 − 3	7 − 6
5 − 3 *2*	7 − 5	2 − 1	6 − 6 *0*	8 − 4	2 − 2 *0*	13 − 6	15 − 8	2 − 0 *2*	13 − 9
1 − 1 *0*	11 − 9	10 − 4	9 − 2	14 − 6	8 − 0 *8*	9 − 4 *5*	10 − 2	6 − 5 *1*	8 − 3
7 − 7 *0*	14 − 8	12 − 9 *3*	9 − 8	12 − 7	9 − 9 *0*	15 − 7	8 − 8 *0*	14 − 9	9 − 7

B

100 Subtraction Facts
For use with Lesson 15

Name _____

Time _____

Subtract.

16 − 9	7 − 1	18 − 9	11 − 3	13 − 7	8 − 2	11 − 5	5 − 0	17 − 9	6 − 1
10 − 9	6 − 2	13 − 4	4 − 0	10 − 5	5 − 1	10 − 3	12 − 6	10 − 1	6 − 4
7 − 2	14 − 7	8 − 1	11 − 6	3 − 3	16 − 7	5 − 2	12 − 4	3 − 0	11 − 7
17 − 8	6 − 0	10 − 6	4 − 1	9 − 5	9 − 0	5 − 4	12 − 5	4 − 2	9 − 3
12 − 3	16 − 8	9 − 1	15 − 6	11 − 4	13 − 5	1 − 0	8 − 5	9 − 6	11 − 2
7 − 0	10 − 8	6 − 3	14 − 5	3 − 1	8 − 6	4 − 4	11 − 8	3 − 2	15 − 9
13 − 8	7 − 4	10 − 7	0 − 0	12 − 8	5 − 5	4 − 3	8 − 7	7 − 3	7 − 6
5 − 3	7 − 5	2 − 1	6 − 6	8 − 4	2 − 2	13 − 6	15 − 8	2 − 0	13 − 9
1 − 1	11 − 9	10 − 4	9 − 2	14 − 6	8 − 0	9 − 4	10 − 2	6 − 5	8 − 3
7 − 7	14 − 8	12 − 9	9 − 8	12 − 7	9 − 9	15 − 7	8 − 8	14 − 9	9 − 7

12 | Multiplication Table
For use with Lesson 15

Name _____

	0	1	2	3	4	5	6	7	8	9	10
0											
1											
2											
3											
4											
5											
6											
7											
8											
9											
10											

B 100 Subtraction Facts
For use with Test 2

Name _____

Time _____

Subtract.

16 − 9	7 − 1 **6**	18 − 9 **9**	11 − 3 **8**	13 − 7	8 − 2 **6**	11 − 5 **6**	5 − 0 **5**	17 − 9	6 − 1 **5**
10 − 9 **1**	6 − 2 **4**	13 − 4 **9**	4 − 0 **4**	10 − 5 **5**	5 − 1 **4**	10 − 3	12 − 6 **6**	10 − 1 **(9)**	6 − 4
7 − 2 **5**	14 − 7	8 − 1 **7**	11 − 6 **5**	3 − 3 **0**	16 − 7	5 − 2	12 − 4	3 − 0 **3**	11 − 7
17 − 8	6 − 0 **6**	10 − 6 **4**	4 − 1 **3**	9 − 5 **4**	9 − 0 **9**	5 − 4 **1**	12 − 5	4 − 2	9 − 3
12 − 3 **9**	16 − 8	9 − 1 **8**	15 − 6 **9**	11 − 4	13 − 5	1 − 0 **1**	8 − 5	9 − 6 **3**	11 − 2
7 − 0 **7**	10 − 8 **2**	6 − 3 **3**	14 − 5 **9**	3 − 1 **2**	8 − 6 **S**	4 − 4 **0**	11 − 8	3 − 2 **1**	15 − 9 **6**
13 − 8	7 − 4	10 − 7 **3**	0 − 0 **0**	12 − 8	5 − 5 **0**	4 − 3 **1**	8 − 7	7 − 3	7 − 6
5 − 3 **2**	7 − 5 **2**	2 − 1 **1**	6 − 6 **0**	8 − 4	2 − 2 **0**	13 − 6	15 − 8	2 − 0 **2**	13 − 9
1 − 1 **0**	11 − 9 **2**	10 − 4 **2**	9 − 2 **7**	14 − 6	8 − 0 **8**	9 − 4 **5**	10 − 2 **8**	6 − 5 **1**	8 − 3 **5**
7 − 7 **0**	14 − 8	12 − 9 **3**	9 − 8 **1**	12 − 7 **5**	9 − 9 **0**	15 − 7	8 − 8 **0**	14 − 9	9 − 7

C | **100 Multiplication Facts**
For use with Lesson 16

Name _____

Time _____

Multiply.

9 × 9	3 × 5	8 × 5	2 × 6	4 × 7	0 × 3	7 × 2	1 × 5	7 × 8	4 × 0
3 × 4	5 × 9	0 × 2	7 × 3	4 × 1	2 × 7	6 × 3	5 × 4	1 × 0	9 × 2
1 × 1	9 × 0	2 × 8	6 × 4	0 × 7	8 × 1	3 × 3	4 × 8	9 × 3	2 × 0
4 × 9	7 × 0	1 × 2	8 × 4	6 × 5	2 × 9	9 × 4	0 × 1	7 × 4	5 × 8
0 × 8	4 × 2	9 × 8	3 × 6	5 × 5	1 × 6	5 × 0	6 × 6	2 × 1	7 × 9
9 × 1	2 × 2	5 × 1	4 × 3	0 × 0	8 × 9	3 × 7	9 × 7	1 × 7	6 × 0
5 × 6	7 × 5	3 × 0	8 × 8	1 × 3	8 × 3	5 × 2	0 × 4	9 × 5	6 × 7
2 × 3	8 × 6	0 × 5	6 × 1	3 × 8	7 × 6	1 × 8	9 × 6	4 × 4	5 × 3
7 × 7	1 × 4	6 × 2	4 × 5	2 × 4	8 × 0	3 × 1	6 × 8	0 × 9	8 × 7
3 × 2	4 × 6	1 × 9	5 × 7	8 × 2	0 × 6	7 × 1	2 × 5	6 × 9	3 × 9

C

100 Multiplication Facts
For use with Lesson 17

Name _____

Time _____

Multiply.

9 ×9	3 ×5	8 ×5	2 ×6	4 ×7	0 ×3 *0*	7 ×2	1 ×5	7 ×8	4 ×0 *0*
3 ×4	5 ×9	0 ×2 *0*	7 ×3	4 ×1	2 ×7	6 ×3	5 ×4	1 ×0 *0*	9 ×2
1 ×1	9 ×0 *0*	2 ×8	6 ×4	0 ×7 *0*	8 ×1 *8*	3 ×3	4 ×8	9 ×3	2 ×0 *0*
4 ×9	7 ×0 *0*	1 ×2	8 ×4	6 ×5	2 ×9 *18*	9 ×4	0 ×1 *0*	7 ×4	5 ×8
0 ×8 *0*	4 ×2	9 ×8	3 ×6	5 ×5	1 ×6	5 ×0 *0*	6 ×6	2 ×1 *2*	7 ×9
9 ×1	2 ×2 *4*	5 ×1 *5*	4 ×3 *12*	0 ×0 *0*	8 ×9	3 ×7	9 ×7	1 ×7 *7*	6 ×0 *0*
5 ×6 *30*	7 ×5	3 ×0	8 ×8	1 ×3	8 ×3	5 ×2	0 ×4	9 ×5	6 ×7
2 ×3	8 ×6	0 ×5	6 ×1	3 ×8	7 ×6	1 ×8 *8*	9 ×6	4 ×4	5 ×3
7 ×7	1 ×4 *4*	6 ×2	4 ×5	2 ×4	8 ×0 *8*	3 ×1 *3*	6 ×8	0 ×9 *0*	8 ×7
3 ×2	4 ×6	1 ×9	5 ×7	8 ×2	0 ×6	7 ×1	2 ×5 *10*	6 ×9	3 ×9

C | **100 Multiplication Facts**
For use with Lesson 18

Name _____
Time _____

Multiply.

9 × 9	3 × 5	8 × 5	2 × 6	4 × 7	0 × 3	7 × 2	1 × 5	7 × 8	4 × 0
3 × 4	5 × 9	0 × 2	7 × 3	4 × 1	2 × 7	6 × 3	5 × 4	1 × 0	9 × 2
1 × 1	9 × 0	2 × 8	6 × 4	0 × 7	8 × 1	3 × 3	4 × 8	9 × 3	2 × 0
4 × 9	7 × 0	1 × 2	8 × 4	6 × 5	2 × 9	9 × 4	0 × 1	7 × 4	5 × 8
0 × 8	4 × 2	9 × 8	3 × 6	5 × 5	1 × 6	5 × 0	6 × 6	2 × 1	7 × 9
9 × 1	2 × 2	5 × 1	4 × 3	0 × 0	8 × 9	3 × 7	9 × 7	1 × 7	6 × 0
5 × 6	7 × 5	3 × 0	8 × 8	1 × 3	8 × 3	5 × 2	0 × 4	9 × 5	6 × 7
2 × 3	8 × 6	0 × 5	6 × 1	3 × 8	7 × 6	1 × 8	9 × 6	4 × 4	5 × 3
7 × 7	1 × 4	6 × 2	4 × 5	2 × 4	8 × 0	3 × 1	6 × 8	0 × 9	8 × 7
3 × 2	4 × 6	1 × 9	5 × 7	8 × 2	0 × 6	7 × 1	2 × 5	6 × 9	3 × 9

C **100 Multiplication Facts**
For use with Lesson 19

Name _____

Time _____

Multiply.

9 × 9	3 × 5	8 × 5	2 × 6	4 × 7	0 × 3	7 × 2	1 × 5	7 × 8	4 × 0
3 × 4	5 × 9	0 × 2	7 × 3	4 × 1	2 × 7	6 × 3	5 × 4	1 × 0	9 × 2
1 × 1	9 × 0	2 × 8	6 × 4	0 × 7	8 × 1	3 × 3	4 × 8	9 × 3	2 × 0
4 × 9	7 × 0	1 × 2	8 × 4	6 × 5	2 × 9	9 × 4	0 × 1	7 × 4	5 × 8
0 × 8	4 × 2	9 × 8	3 × 6	5 × 5	1 × 6	5 × 0	6 × 6	2 × 1	7 × 9
9 × 1	2 × 2	5 × 1	4 × 3	0 × 0	8 × 9	3 × 7	9 × 7	1 × 7	6 × 0
5 × 6	7 × 5	3 × 0	8 × 8	1 × 3	8 × 3	5 × 2	0 × 4	9 × 5	6 × 7
2 × 3	8 × 6	0 × 5	6 × 1	3 × 8	7 × 6	1 × 8	9 × 6	4 × 4	5 × 3
7 × 7	1 × 4	6 × 2	4 × 5	2 × 4	8 × 0	3 × 1	6 × 8	0 × 9	8 × 7
3 × 2	4 × 6	1 × 9	5 × 7	8 × 2	0 × 6	7 × 1	2 × 5	6 × 9	3 × 9

D	**90 Division Facts** *For use with Lesson 20*

Name _____

Time _____

Divide.

$7\overline{)21}$	$2\overline{)10}$	$6\overline{)42}$	$1\overline{)3}$	$4\overline{)24}$	$3\overline{)6}$	$9\overline{)54}$	$6\overline{)18}$	$4\overline{)0}$	$5\overline{)30}$
$4\overline{)32}$	$8\overline{)56}$	$1\overline{)0}$	$6\overline{)12}$	$3\overline{)18}$	$9\overline{)72}$	$5\overline{)15}$	$2\overline{)8}$	$7\overline{)42}$	$6\overline{)36}$
$6\overline{)0}$	$5\overline{)10}$	$9\overline{)9}$	$2\overline{)6}$	$7\overline{)63}$	$4\overline{)16}$	$8\overline{)48}$	$1\overline{)2}$	$5\overline{)35}$	$3\overline{)21}$
$2\overline{)18}$	$6\overline{)6}$	$3\overline{)15}$	$8\overline{)40}$	$2\overline{)0}$	$5\overline{)20}$	$9\overline{)27}$	$1\overline{)8}$	$4\overline{)4}$	$7\overline{)35}$
$4\overline{)20}$	$9\overline{)63}$	$1\overline{)4}$	$7\overline{)14}$	$3\overline{)3}$	$8\overline{)24}$	$5\overline{)0}$	$6\overline{)24}$	$8\overline{)8}$	$2\overline{)16}$
$5\overline{)5}$	$8\overline{)64}$	$3\overline{)0}$	$4\overline{)28}$	$7\overline{)49}$	$2\overline{)4}$	$9\overline{)81}$	$3\overline{)12}$	$6\overline{)30}$	$1\overline{)5}$
$8\overline{)32}$	$1\overline{)1}$	$9\overline{)36}$	$3\overline{)27}$	$2\overline{)14}$	$5\overline{)25}$	$6\overline{)48}$	$8\overline{)0}$	$7\overline{)28}$	$4\overline{)36}$
$2\overline{)12}$	$5\overline{)45}$	$1\overline{)7}$	$4\overline{)8}$	$7\overline{)0}$	$8\overline{)16}$	$3\overline{)24}$	$9\overline{)45}$	$1\overline{)9}$	$6\overline{)54}$
$7\overline{)56}$	$9\overline{)0}$	$8\overline{)72}$	$2\overline{)2}$	$5\overline{)40}$	$3\overline{)9}$	$9\overline{)18}$	$1\overline{)6}$	$4\overline{)12}$	$7\overline{)7}$

C

100 Multiplication Facts
For use with Test 3

Name _____

Time _____

Multiply.

9 × 9	3 × 5	8 × 5	2 × 6 **15**	4 × 7	0 × 3	7 × 2	1 × 5	7 × 8	4 × 0
3 × 4	5 × 9	0 × 2	7 × 3	4 × 1 **4**	2 × 7	6 × 3	5 × 4	1 × 0	9 × 2
1 × 1	9 × 0	2 × 8	6 × 4	0 × 7	8 × 1 **8**	3 × 3	4 × 8	9 × 3	2 × 0
4 × 9	7 × 0	1 × 2	8 × 4	6 × 5	2 × 9	9 × 4	0 × 1	7 × 4	5 × 8
0 × 8	4 × 2	9 × 8	3 × 6	5 × 5 **25**	1 × 6	5 × 0	6 × 6	2 × 1	7 × 9
9 × 1	2 × 2	5 × 1	4 × 3	0 × 0 **0**	8 × 9	3 × 7	9 × 7	1 × 7	6 × 0
5 × 6	7 × 5	3 × 0	8 × 8	1 × 3 **3**	8 × 3	5 × 2	0 × 4	9 × 5	6 × 7
2 × 3	8 × 6	0 × 5 **0**	6 × 1 **6**	3 × 8 **54**	7 × 6	1 × 8	9 × 6	4 × 4	5 × 3
7 × 7	1 × 4 **4**	6 × 2	4 × 5	2 × 4 **8**	8 × 0	3 × 1	6 × 8	0 × 9	8 × 7
3 × 2 **6**	4 × 6	1 × 9 **9**	5 × 7	8 × 2	0 × 6	7 × 1	2 × 5	6 × 9	3 × 9

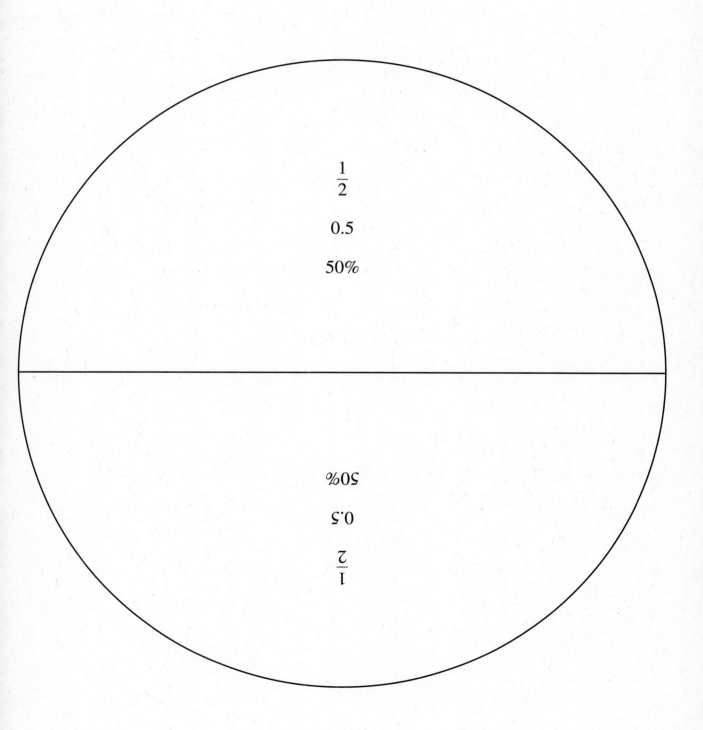

14 | Fourths
For use with Investigation 2

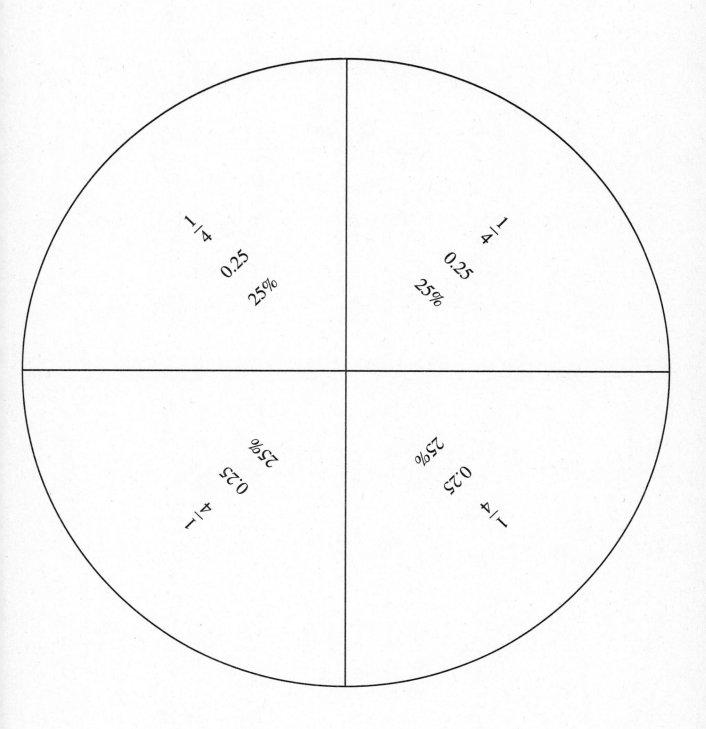

15 Tenths

For use with Investigation 2

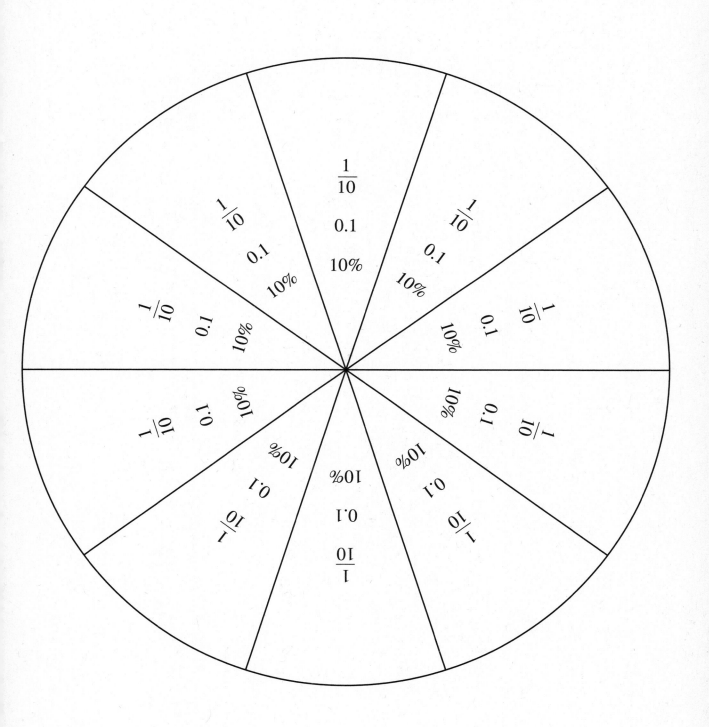

E 90 Division Facts
For use with Lesson 21

Name _____

Time _____

Divide.

20 ÷ 4 =	21 ÷ 7 =	0 ÷ 2 =	27 ÷ 3 =	8 ÷ 1 =	54 ÷ 6 =
15 ÷ 5 =	6 ÷ 3 =	28 ÷ 4 =	18 ÷ 2 =	24 ÷ 6 =	9 ÷ 9 =
56 ÷ 8 =	0 ÷ 6 =	21 ÷ 3 =	1 ÷ 1 =	25 ÷ 5 =	12 ÷ 2 =
5 ÷ 1 =	45 ÷ 9 =	16 ÷ 4 =	30 ÷ 6 =	9 ÷ 3 =	14 ÷ 7 =
0 ÷ 8 =	6 ÷ 2 =	24 ÷ 8 =	10 ÷ 5 =	81 ÷ 9 =	24 ÷ 4 =
16 ÷ 2 =	30 ÷ 5 =	0 ÷ 1 =	28 ÷ 7 =	4 ÷ 4 =	40 ÷ 8 =
3 ÷ 3 =	18 ÷ 6 =	63 ÷ 9 =	40 ÷ 5 =	10 ÷ 2 =	36 ÷ 6 =
32 ÷ 8 =	12 ÷ 4 =	18 ÷ 3 =	35 ÷ 7 =	8 ÷ 8 =	2 ÷ 1 =
45 ÷ 5 =	7 ÷ 7 =	27 ÷ 9 =	9 ÷ 1 =	48 ÷ 6 =	0 ÷ 7 =
4 ÷ 1 =	0 ÷ 9 =	24 ÷ 3 =	32 ÷ 4 =	5 ÷ 5 =	72 ÷ 9 =
56 ÷ 7 =	15 ÷ 3 =	12 ÷ 6 =	8 ÷ 2 =	63 ÷ 7 =	0 ÷ 4 =
14 ÷ 2 =	42 ÷ 6 =	6 ÷ 1 =	16 ÷ 8 =	20 ÷ 5 =	49 ÷ 7 =
36 ÷ 4 =	64 ÷ 8 =	0 ÷ 3 =	54 ÷ 9 =	4 ÷ 2 =	48 ÷ 8 =
18 ÷ 9 =	3 ÷ 1 =	35 ÷ 5 =	8 ÷ 4 =	72 ÷ 8 =	6 ÷ 6 =
0 ÷ 5 =	42 ÷ 7 =	2 ÷ 2 =	36 ÷ 9 =	7 ÷ 1 =	12 ÷ 3 =

F | 64 Multiplication Facts
For use with Lesson 22

Name _____

Time _____

Multiply.

5 × 6	4 × 3	9 × 8	7 × 5	2 × 9	8 × 4	9 × 3	6 × 9
9 × 4	2 × 5	7 × 6	4 × 8	7 × 9	5 × 4	3 × 2	9 × 7
3 × 7	8 × 5	6 × 2	5 × 5	3 × 5	2 × 4	7 × 7	8 × 9
6 × 4	2 × 8	4 × 4	8 × 2	3 × 9	6 × 6	9 × 9	5 × 3
4 × 6	8 × 8	5 × 7	6 × 3	2 × 2	7 × 4	3 × 8	8 × 6
2 × 6	5 × 9	3 × 3	9 × 2	6 × 7	4 × 5	7 × 2	9 × 6
5 × 2	7 × 8	2 × 3	6 × 8	4 × 7	9 × 5	3 × 6	8 × 7
3 × 4	7 × 3	5 × 8	4 × 2	8 × 3	2 × 7	6 × 5	4 × 9

D   90 Division Facts

For use with Lesson 23

Name _____

Time _____

Divide.

$7\overline{)21}$	$2\overline{)10}$	$6\overline{)42}$	$1\overline{)3}$	$4\overline{)24}$	$3\overline{)6}$	$9\overline{)54}$	$6\overline{)18}$	$4\overline{)0}$	$5\overline{)30}$
$4\overline{)32}$	$8\overline{)56}$	$1\overline{)0}$	$6\overline{)12}$	$3\overline{)18}$	$9\overline{)72}$	$5\overline{)15}$	$2\overline{)8}$	$7\overline{)42}$	$6\overline{)36}$
$6\overline{)0}$	$5\overline{)10}$	$9\overline{)9}$	$2\overline{)6}$	$7\overline{)63}$	$4\overline{)16}$	$8\overline{)48}$	$1\overline{)2}$	$5\overline{)35}$	$3\overline{)21}$
$2\overline{)18}$	$6\overline{)6}$	$3\overline{)15}$	$8\overline{)40}$	$2\overline{)0}$	$5\overline{)20}$	$9\overline{)27}$	$1\overline{)8}$	$4\overline{)4}$	$7\overline{)35}$
$4\overline{)20}$	$9\overline{)63}$	$1\overline{)4}$	$7\overline{)14}$	$3\overline{)3}$	$8\overline{)24}$	$5\overline{)0}$	$6\overline{)24}$	$8\overline{)8}$	$2\overline{)16}$
$5\overline{)5}$	$8\overline{)64}$	$3\overline{)0}$	$4\overline{)28}$	$7\overline{)49}$	$2\overline{)4}$	$9\overline{)81}$	$3\overline{)12}$	$6\overline{)30}$	$1\overline{)5}$
$8\overline{)32}$	$1\overline{)1}$	$9\overline{)36}$	$3\overline{)27}$	$2\overline{)14}$	$5\overline{)25}$	$6\overline{)48}$	$8\overline{)0}$	$7\overline{)28}$	$4\overline{)36}$
$2\overline{)12}$	$5\overline{)45}$	$1\overline{)7}$	$4\overline{)8}$	$7\overline{)0}$	$8\overline{)16}$	$3\overline{)24}$	$9\overline{)45}$	$1\overline{)9}$	$6\overline{)54}$
$7\overline{)56}$	$9\overline{)0}$	$8\overline{)72}$	$2\overline{)2}$	$5\overline{)40}$	$3\overline{)9}$	$9\overline{)18}$	$1\overline{)6}$	$4\overline{)12}$	$7\overline{)7}$

| F |

64 Multiplication Facts
For use with Lesson 24

Name _____

Time _____

Multiply.

5 × 6 **30**	4 × 3 **12**	9 × 8	7 × 5	2 × 9	8 × 4	9 × 3	6 × 9
9 × 4	2 × 5 **10**	7 × 6	4 × 8	7 × 9	5 × 4	3 × 2 **6**	9 × 7
3 × 7	8 × 5	6 × 2	5 × 5 **25**	3 × 5 **15**	2 × 4	7 × 7	8 × 9
6 × 4	2 × 8 **16**	4 × 4 **16**	8 × 2	3 × 9	6 × 6	9 × 9	5 × 3
4 × 6	8 × 8	5 × 7	6 × 3	2 × 2 **4**	7 × 4	3 × 8	8 × 6
2 × 6	5 × 9	3 × 3	9 × 2	6 × 7	4 × 5 **20**	7 × 2	9 × 6
5 × 2 **10**	7 × 8	2 × 3 **6**	6 × 8	4 × 7	9 × 5	3 × 6	8 × 7
3 × 4 **12**	7 × 3	5 × 8	4 × 2	8 × 3	2 × 7	6 × 5 **30**	4 × 9

E	**90 Division Facts**	Name _____
	For use with Lesson 25	Time _____

Divide.

20 ÷ 4 =	21 ÷ 7 =	0 ÷ 2 =	27 ÷ 3 =	8 ÷ 1 =	54 ÷ 6 =
15 ÷ 5 =	6 ÷ 3 =	28 ÷ 4 =	18 ÷ 2 =	24 ÷ 6 =	9 ÷ 9 =
56 ÷ 8 =	0 ÷ 6 =	21 ÷ 3 =	1 ÷ 1 =	25 ÷ 5 =	12 ÷ 2 =
5 ÷ 1 =	45 ÷ 9 =	16 ÷ 4 =	30 ÷ 6 =	9 ÷ 3 =	14 ÷ 7 =
0 ÷ 8 =	6 ÷ 2 =	24 ÷ 8 =	10 ÷ 5 =	81 ÷ 9 =	24 ÷ 4 =
16 ÷ 2 =	30 ÷ 5 =	0 ÷ 1 =	28 ÷ 7 =	4 ÷ 4 =	40 ÷ 8 =
3 ÷ 3 =	18 ÷ 6 =	63 ÷ 9 =	40 ÷ 5 =	10 ÷ 2 =	36 ÷ 6 =
32 ÷ 8 =	12 ÷ 4 =	18 ÷ 3 =	35 ÷ 7 =	8 ÷ 8 =	2 ÷ 1 =
45 ÷ 5 =	7 ÷ 7 =	27 ÷ 9 =	9 ÷ 1 =	48 ÷ 6 =	0 ÷ 7 =
4 ÷ 1 =	0 ÷ 9 =	24 ÷ 3 =	32 ÷ 4 =	5 ÷ 5 =	72 ÷ 9 =
56 ÷ 7 =	15 ÷ 3 =	12 ÷ 6 =	8 ÷ 2 =	63 ÷ 7 =	0 ÷ 4 =
14 ÷ 2 =	42 ÷ 6 =	6 ÷ 1 =	16 ÷ 8 =	20 ÷ 5 =	49 ÷ 7 =
36 ÷ 4 =	64 ÷ 8 =	0 ÷ 3 =	54 ÷ 9 =	4 ÷ 2 =	48 ÷ 8 =
18 ÷ 9 =	3 ÷ 1 =	35 ÷ 5 =	8 ÷ 4 =	72 ÷ 8 =	6 ÷ 6 =
0 ÷ 5 =	42 ÷ 7 =	2 ÷ 2 =	36 ÷ 9 =	7 ÷ 1 =	12 ÷ 3 =

F   64 Multiplication Facts
For use with Test 4

Name _____

Time _____

Multiply.

5 × 6	4 × 3	9 × 8	7 × 5	2 × 9	8 × 4	9 × 3	6 × 9
9 × 4	2 × 5	7 × 6	4 × 8	7 × 9	5 × 4	3 × 2	9 × 7
3 × 7	8 × 5	6 × 2	5 × 5	3 × 5	2 × 4	7 × 7	8 × 9
6 × 4	2 × 8	4 × 4	8 × 2	3 × 9	6 × 6	9 × 9	5 × 3
4 × 6	8 × 8	5 × 7	6 × 3	2 × 2	7 × 4	3 × 8	8 × 6
2 × 6	5 × 9	3 × 3	9 × 2	6 × 7	4 × 5	7 × 2	9 × 6
5 × 2	7 × 8	2 × 3	6 × 8	4 × 7	9 × 5	3 × 6	8 × 7
3 × 4	7 × 3	5 × 8	4 × 2	8 × 3	2 × 7	6 × 5	4 × 9

F 64 Multiplication Facts

For use with Lesson 26

Name _____

Time _____

Multiply.

5 × 6	4 × 3	9 × 8	7 × 5	2 × 9	8 × 4	9 × 3	6 × 9
9 × 4	2 × 5	7 × 6	4 × 8	7 × 9	5 × 4	3 × 2	9 × 7
3 × 7	8 × 5	6 × 2	5 × 5	3 × 5	2 × 4	7 × 7	8 × 9
6 × 4	2 × 8	4 × 4	8 × 2	3 × 9	6 × 6	9 × 9	5 × 3
4 × 6	8 × 8	5 × 7	6 × 3	2 × 2	7 × 4	3 × 8	8 × 6
2 × 6	5 × 9	3 × 3	9 × 2	6 × 7	4 × 5	7 × 2	9 × 6
5 × 2	7 × 8	2 × 3	6 × 8	4 × 7	9 × 5	3 × 6	8 × 7
3 × 4	7 × 3	5 × 8	4 × 2	8 × 3	2 × 7	6 × 5	4 × 9

F	**64 Multiplication Facts**	Name _____
	For use with Lesson 27	Time _____

Multiply.

5 × 6	4 × 3	9 × 8	7 × 5	2 × 9	8 × 4	9 × 3	6 × 9
9 × 4	2 × 5	7 × 6	4 × 8	7 × 9	5 × 4	3 × 2	9 × 7
3 × 7	8 × 5	6 × 2	5 × 5	3 × 5	2 × 4	7 × 7	8 × 9
6 × 4	2 × 8	4 × 4	8 × 2	3 × 9	6 × 6	9 × 9	5 × 3
4 × 6	8 × 8	5 × 7	6 × 3	2 × 2	7 × 4	3 × 8	8 × 6
2 × 6	5 × 9	3 × 3	9 × 2	6 × 7	4 × 5	7 × 2	9 × 6
5 × 2	7 × 8	2 × 3	6 × 8	4 × 7	9 × 5	3 × 6	8 × 7
3 × 4	7 × 3	5 × 8	4 × 2	8 × 3	2 × 7	6 × 5	4 × 9

FACTS PRACTICE TEST

90 Division Facts
For use with Lesson 28

Name _____

Time _____

Divide.

7)21	2)10	6)42	1)3	4)24	3)6	9)54	6)18	4)0	5)30
4)32	8)56	1)0	6)12	3)18	9)72	5)15	2)8	7)42	6)36
6)0	5)10	9)9	2)6	7)63	4)16	8)48	1)2	5)35	3)21
2)18	6)6	3)15	8)40	2)0	5)20	9)27	1)8	4)4	7)35
4)20	9)63	1)4	7)14	3)3	8)24	5)0	6)24	8)8	2)16
5)5	8)64	3)0	4)28	7)49	2)4	9)81	3)12	6)30	1)5
8)32	1)1	9)36	3)27	2)14	5)25	6)48	8)0	7)28	4)36
2)12	5)45	1)7	4)8	7)0	8)16	3)24	9)45	1)9	6)54
7)56	9)0	8)72	2)2	5)40	3)9	9)18	1)6	4)12	7)7

Saxon Math 6/5—Homeschool

F

64 Multiplication Facts
For use with Lesson 29

Name _____

Time _____

Multiply.

5 × 6	4 × 3	9 × 8	7 × 5	2 × 9	8 × 4	9 × 3	6 × 9
9 × 4	2 × 5	7 × 6	4 × 8	7 × 9	5 × 4	3 × 2	9 × 7
3 × 7	8 × 5	6 × 2	5 × 5	3 × 5	2 × 4	7 × 7	8 × 9
6 × 4	2 × 8	4 × 4	8 × 2	3 × 9	6 × 6	9 × 9	5 × 3
4 × 6	8 × 8	5 × 7	6 × 3	2 × 2	7 × 4	3 × 8	8 × 6
2 × 6	5 × 9	3 × 3	9 × 2	6 × 7	4 × 5	7 × 2	9 × 6
5 × 2	7 × 8	2 × 3	6 × 8	4 × 7	9 × 5	3 × 6	8 × 7
3 × 4	7 × 3	5 × 8	4 × 2	8 × 3	2 × 7	6 × 5	4 × 9

E	**90 Division Facts**
	For use with Lesson 30

Name _____

Time _____

Divide.

20 ÷ 4 =	21 ÷ 7 =	0 ÷ 2 =	27 ÷ 3 =	8 ÷ 1 =	54 ÷ 6 =
15 ÷ 5 =	6 ÷ 3 =	28 ÷ 4 =	18 ÷ 2 =	24 ÷ 6 =	9 ÷ 9 =
56 ÷ 8 =	0 ÷ 6 =	21 ÷ 3 =	1 ÷ 1 =	25 ÷ 5 =	12 ÷ 2 =
5 ÷ 1 =	45 ÷ 9 =	16 ÷ 4 =	30 ÷ 6 =	9 ÷ 3 =	14 ÷ 7 =
0 ÷ 8 =	6 ÷ 2 =	24 ÷ 8 =	10 ÷ 5 =	81 ÷ 9 =	24 ÷ 4 =
16 ÷ 2 =	30 ÷ 5 =	0 ÷ 1 =	28 ÷ 7 =	4 ÷ 4 =	40 ÷ 8 =
3 ÷ 3 =	18 ÷ 6 =	63 ÷ 9 =	40 ÷ 5 =	10 ÷ 2 =	36 ÷ 6 =
32 ÷ 8 =	12 ÷ 4 =	18 ÷ 3 =	35 ÷ 7 =	8 ÷ 8 =	2 ÷ 1 =
45 ÷ 5 =	7 ÷ 7 =	27 ÷ 9 =	9 ÷ 1 =	48 ÷ 6 =	0 ÷ 7 =
4 ÷ 1 =	0 ÷ 9 =	24 ÷ 3 =	32 ÷ 4 =	5 ÷ 5 =	72 ÷ 9 =
56 ÷ 7 =	15 ÷ 3 =	12 ÷ 6 =	8 ÷ 2 =	63 ÷ 7 =	0 ÷ 4 =
14 ÷ 2 =	42 ÷ 6 =	6 ÷ 1 =	16 ÷ 8 =	20 ÷ 5 =	49 ÷ 7 =
36 ÷ 4 =	64 ÷ 8 =	0 ÷ 3 =	54 ÷ 9 =	4 ÷ 2 =	48 ÷ 8 =
18 ÷ 9 =	3 ÷ 1 =	35 ÷ 5 =	8 ÷ 4 =	72 ÷ 8 =	6 ÷ 6 =
0 ÷ 5 =	42 ÷ 7 =	2 ÷ 2 =	36 ÷ 9 =	7 ÷ 1 =	12 ÷ 3 =

F	**64 Multiplication Facts** *For use with Test 5*

Name _____

Time _____

Multiply.

5 × 6	4 × 3	9 × 8	7 × 5	2 × 9	8 × 4	9 × 3	6 × 9
9 × 4	2 × 5	7 × 6	4 × 8	7 × 9	5 × 4	3 × 2	9 × 7
3 × 7	8 × 5	6 × 2	5 × 5	3 × 5	2 × 4	7 × 7	8 × 9
6 × 4	2 × 8	4 × 4	8 × 2	3 × 9	6 × 6	9 × 9	5 × 3
4 × 6	8 × 8	5 × 7	6 × 3	2 × 2	7 × 4	3 × 8	8 × 6
2 × 6	5 × 9	3 × 3	9 × 2	6 × 7	4 × 5	7 × 2	9 × 6
5 × 2	7 × 8	2 × 3	6 × 8	4 × 7	9 × 5	3 × 6	8 × 7
3 × 4	7 × 3	5 × 8	4 × 2	8 × 3	2 × 7	6 × 5	4 × 9

16 | Thirds
For use with Investigation 3

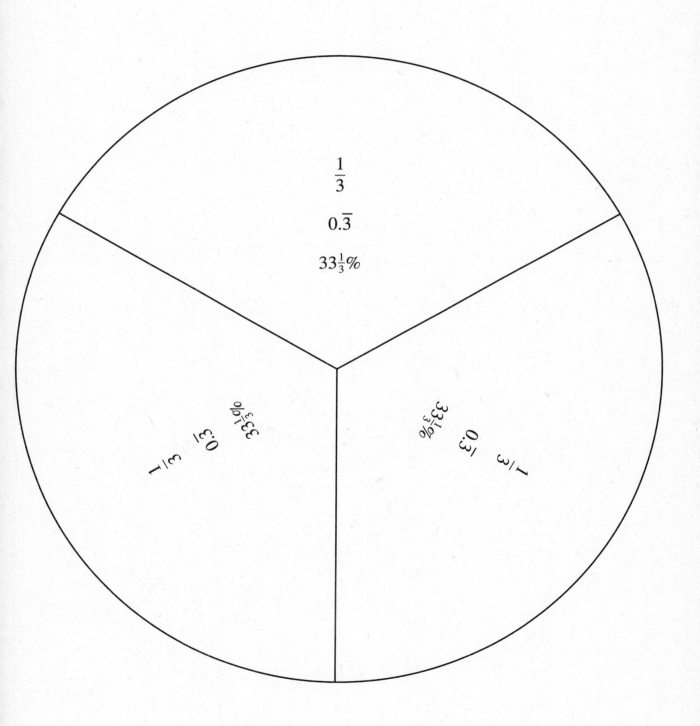

$\frac{1}{3}$

$0.\overline{3}$

$33\frac{1}{3}\%$

17 | Fifths
For use with Investigation 3

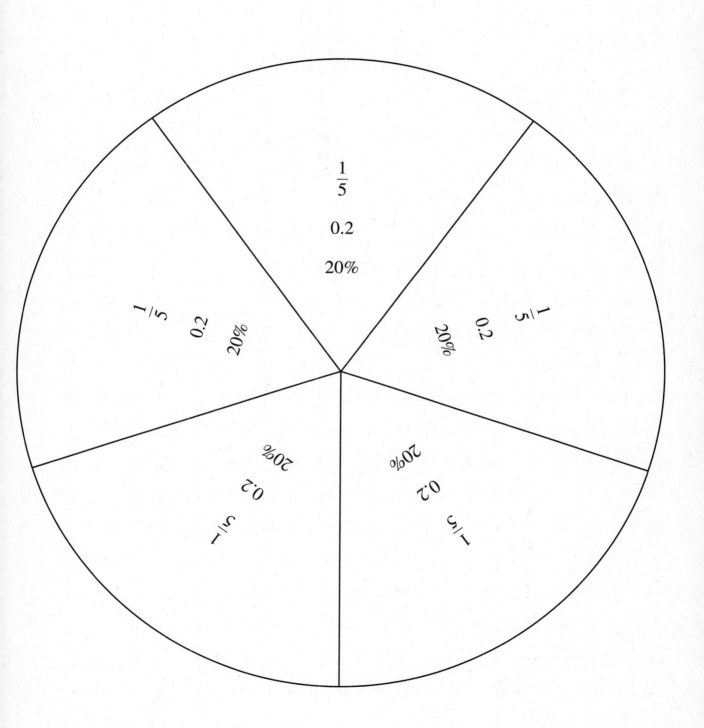

18 | Eighths

For use with Investigation 3

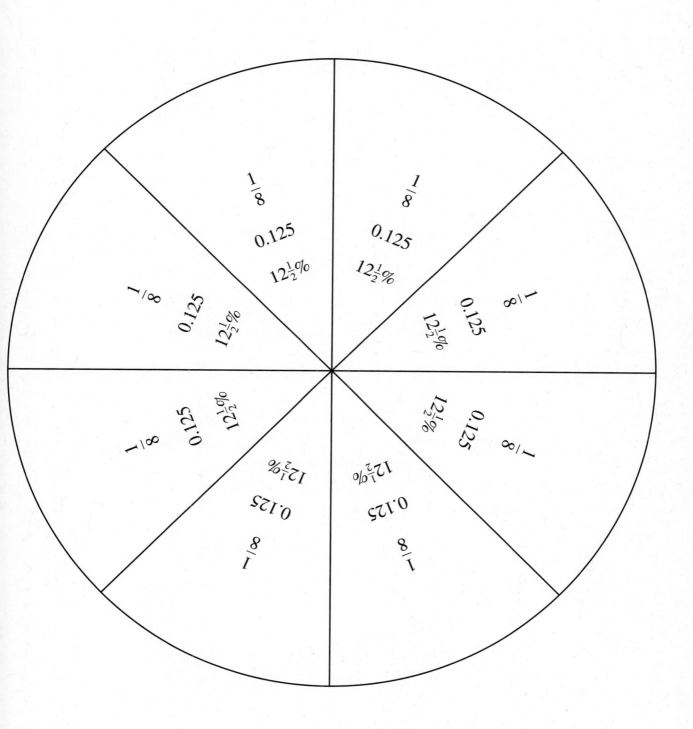

F

64 Multiplication Facts
For use with Lesson 31

Name _____

Time _____

Multiply.

5 × 6	4 × 3	9 × 8	7 × 5	2 × 9	8 × 4	9 × 3	6 × 9
9 × 4	2 × 5	7 × 6	4 × 8	7 × 9	5 × 4	3 × 2	9 × 7
3 × 7	8 × 5	6 × 2	5 × 5	3 × 5	2 × 4	7 × 7	8 × 9
6 × 4	2 × 8	4 × 4	8 × 2	3 × 9	6 × 6	9 × 9	5 × 3
4 × 6	8 × 8	5 × 7	6 × 3	2 × 2	7 × 4	3 × 8	8 × 6
2 × 6	5 × 9	3 × 3	9 × 2	6 × 7	4 × 5	7 × 2	9 × 6
5 × 2	7 × 8	2 × 3	6 × 8	4 × 7	9 × 5	3 × 6	8 × 7
3 × 4	7 × 3	5 × 8	4 × 2	8 × 3	2 × 7	6 × 5	4 × 9

D	**90 Division Facts**	Name _____
	For use with Lesson 32	Time _____

Divide.

7)21	2)10	6)42	1)3	4)24	3)6	9)54	6)18	4)0	5)30
4)32	8)56	1)0	6)12	3)18	9)72	5)15	2)8	7)42	6)36
6)0	5)10	9)9	2)6	7)63	4)16	8)48	1)2	5)35	3)21
2)18	6)6	3)15	8)40	2)0	5)20	9)27	1)8	4)4	7)35
4)20	9)63	1)4	7)14	3)3	8)24	5)0	6)24	8)8	2)16
5)5	8)64	3)0	4)28	7)49	2)4	9)81	3)12	6)30	1)5
8)32	1)1	9)36	3)27	2)14	5)25	6)48	8)0	7)28	4)36
2)12	5)45	1)7	4)8	7)0	8)16	3)24	9)45	1)9	6)54
7)56	9)0	8)72	2)2	5)40	3)9	9)18	1)6	4)12	7)7

F	**64 Multiplication Facts**
	For use with Lesson 33

Name _____

Time _____

Multiply.

5 × 6	4 × 3	9 × 8	7 × 5	2 × 9	8 × 4	9 × 3	6 × 9
9 × 4	2 × 5	7 × 6	4 × 8	7 × 9	5 × 4	3 × 2	9 × 7
3 × 7	8 × 5	6 × 2	5 × 5	3 × 5	2 × 4	7 × 7	8 × 9
6 × 4	2 × 8	4 × 4	8 × 2	3 × 9	6 × 6	9 × 9	5 × 3
4 × 6	8 × 8	5 × 7	6 × 3	2 × 2	7 × 4	3 × 8	8 × 6
2 × 6	5 × 9	3 × 3	9 × 2	6 × 7	4 × 5	7 × 2	9 × 6
5 × 2	7 × 8	2 × 3	6 × 8	4 × 7	9 × 5	3 × 6	8 × 7
3 × 4	7 × 3	5 × 8	4 × 2	8 × 3	2 × 7	6 × 5	4 × 9

| | E | **90 Division Facts** | Name _____ |
| | | For use with Lesson 34 | Time _____ |

Divide.

20 ÷ 4 =	21 ÷ 7 =	0 ÷ 2 =	27 ÷ 3 =	8 ÷ 1 =	54 ÷ 6 =
15 ÷ 5 =	6 ÷ 3 =	28 ÷ 4 =	18 ÷ 2 =	24 ÷ 6 =	9 ÷ 9 =
56 ÷ 8 =	0 ÷ 6 =	21 ÷ 3 =	1 ÷ 1 =	25 ÷ 5 =	12 ÷ 2 =
5 ÷ 1 =	45 ÷ 9 =	16 ÷ 4 =	30 ÷ 6 =	9 ÷ 3 =	14 ÷ 7 =
0 ÷ 8 =	6 ÷ 2 =	24 ÷ 8 =	10 ÷ 5 =	81 ÷ 9 =	24 ÷ 4 =
16 ÷ 2 =	30 ÷ 5 =	0 ÷ 1 =	28 ÷ 7 =	4 ÷ 4 =	40 ÷ 8 =
3 ÷ 3 =	18 ÷ 6 =	63 ÷ 9 =	40 ÷ 5 =	10 ÷ 2 =	36 ÷ 6 =
32 ÷ 8 =	12 ÷ 4 =	18 ÷ 3 =	35 ÷ 7 =	8 ÷ 8 =	2 ÷ 1 =
45 ÷ 5 =	7 ÷ 7 =	27 ÷ 9 =	9 ÷ 1 =	48 ÷ 6 =	0 ÷ 7 =
4 ÷ 1 =	0 ÷ 9 =	24 ÷ 3 =	32 ÷ 4 =	5 ÷ 5 =	72 ÷ 9 =
56 ÷ 7 =	15 ÷ 3 =	12 ÷ 6 =	8 ÷ 2 =	63 ÷ 7 =	0 ÷ 4 =
14 ÷ 2 =	42 ÷ 6 =	6 ÷ 1 =	16 ÷ 8 =	20 ÷ 5 =	49 ÷ 7 =
36 ÷ 4 =	64 ÷ 8 =	0 ÷ 3 =	54 ÷ 9 =	4 ÷ 2 =	48 ÷ 8 =
18 ÷ 9 =	3 ÷ 1 =	35 ÷ 5 =	8 ÷ 4 =	72 ÷ 8 =	6 ÷ 6 =
0 ÷ 5 =	42 ÷ 7 =	2 ÷ 2 =	36 ÷ 9 =	7 ÷ 1 =	12 ÷ 3 =

FACTS PRACTICE TEST

F	**64 Multiplication Facts** *For use with Lesson 35*	Name _____ Time _____

Multiply.

5 × 6	4 × 3	9 × 8	7 × 5	2 × 9	8 × 4	9 × 3	6 × 9
9 × 4	2 × 5	7 × 6	4 × 8	7 × 9	5 × 4	3 × 2	9 × 7
3 × 7	8 × 5	6 × 2	5 × 5	3 × 5	2 × 4	7 × 7	8 × 9
6 × 4	2 × 8	4 × 4	8 × 2	3 × 9	6 × 6	9 × 9	5 × 3
4 × 6	8 × 8	5 × 7	6 × 3	2 × 2	7 × 4	3 × 8	8 × 6
2 × 6	5 × 9	3 × 3	9 × 2	6 × 7	4 × 5	7 × 2	9 × 6
5 × 2	7 × 8	2 × 3	6 × 8	4 × 7	9 × 5	3 × 6	8 × 7
3 × 4	7 × 3	5 × 8	4 × 2	8 × 3	2 × 7	6 × 5	4 × 9

D 90 Division Facts

For use with Test 6

Name _____

Time _____

Divide.

$7\overline{)21}$	$2\overline{)10}$	$6\overline{)42}$	$1\overline{)3}$	$4\overline{)24}$	$3\overline{)6}$	$9\overline{)54}$	$6\overline{)18}$	$4\overline{)0}$	$5\overline{)30}$
$4\overline{)32}$	$8\overline{)56}$	$1\overline{)0}$	$6\overline{)12}$	$3\overline{)18}$	$9\overline{)72}$	$5\overline{)15}$	$2\overline{)8}$	$7\overline{)42}$	$6\overline{)36}$
$6\overline{)0}$	$5\overline{)10}$	$9\overline{)9}$	$2\overline{)6}$	$7\overline{)63}$	$4\overline{)16}$	$8\overline{)48}$	$1\overline{)2}$	$5\overline{)35}$	$3\overline{)21}$
$2\overline{)18}$	$6\overline{)6}$	$3\overline{)15}$	$8\overline{)40}$	$2\overline{)0}$	$5\overline{)20}$	$9\overline{)27}$	$1\overline{)8}$	$4\overline{)4}$	$7\overline{)35}$
$4\overline{)20}$	$9\overline{)63}$	$1\overline{)4}$	$7\overline{)14}$	$3\overline{)3}$	$8\overline{)24}$	$5\overline{)0}$	$6\overline{)24}$	$8\overline{)8}$	$2\overline{)16}$
$5\overline{)5}$	$8\overline{)64}$	$3\overline{)0}$	$4\overline{)28}$	$7\overline{)49}$	$2\overline{)4}$	$9\overline{)81}$	$3\overline{)12}$	$6\overline{)30}$	$1\overline{)5}$
$8\overline{)32}$	$1\overline{)1}$	$9\overline{)36}$	$3\overline{)27}$	$2\overline{)14}$	$5\overline{)25}$	$6\overline{)48}$	$8\overline{)0}$	$7\overline{)28}$	$4\overline{)36}$
$2\overline{)12}$	$5\overline{)45}$	$1\overline{)7}$	$4\overline{)8}$	$7\overline{)0}$	$8\overline{)16}$	$3\overline{)24}$	$9\overline{)45}$	$1\overline{)9}$	$6\overline{)54}$
$7\overline{)56}$	$9\overline{)0}$	$8\overline{)72}$	$2\overline{)2}$	$5\overline{)40}$	$3\overline{)9}$	$9\overline{)18}$	$1\overline{)6}$	$4\overline{)12}$	$7\overline{)7}$

D 90 Division Facts

For use with Lesson 36

Name _____

Time _____

Divide.

$7\overline{)21}$	$2\overline{)10}$	$6\overline{)42}$	$1\overline{)3}$	$4\overline{)24}$	$3\overline{)6}$	$9\overline{)54}$	$6\overline{)18}$	$4\overline{)0}$	$5\overline{)30}$
$4\overline{)32}$	$8\overline{)56}$	$1\overline{)0}$	$6\overline{)12}$	$3\overline{)18}$	$9\overline{)72}$	$5\overline{)15}$	$2\overline{)8}$	$7\overline{)42}$	$6\overline{)36}$
$6\overline{)0}$	$5\overline{)10}$	$9\overline{)9}$	$2\overline{)6}$	$7\overline{)63}$	$4\overline{)16}$	$8\overline{)48}$	$1\overline{)2}$	$5\overline{)35}$	$3\overline{)21}$
$2\overline{)18}$	$6\overline{)6}$	$3\overline{)15}$	$8\overline{)40}$	$2\overline{)0}$	$5\overline{)20}$	$9\overline{)27}$	$1\overline{)8}$	$4\overline{)4}$	$7\overline{)35}$
$4\overline{)20}$	$9\overline{)63}$	$1\overline{)4}$	$7\overline{)14}$	$3\overline{)3}$	$8\overline{)24}$	$5\overline{)0}$	$6\overline{)24}$	$8\overline{)8}$	$2\overline{)16}$
$5\overline{)5}$	$8\overline{)64}$	$3\overline{)0}$	$4\overline{)28}$	$7\overline{)49}$	$2\overline{)4}$	$9\overline{)81}$	$3\overline{)12}$	$6\overline{)30}$	$1\overline{)5}$
$8\overline{)32}$	$1\overline{)1}$	$9\overline{)36}$	$3\overline{)27}$	$2\overline{)14}$	$5\overline{)25}$	$6\overline{)48}$	$8\overline{)0}$	$7\overline{)28}$	$4\overline{)36}$
$2\overline{)12}$	$5\overline{)45}$	$1\overline{)7}$	$4\overline{)8}$	$7\overline{)0}$	$8\overline{)16}$	$3\overline{)24}$	$9\overline{)45}$	$1\overline{)9}$	$6\overline{)54}$
$7\overline{)56}$	$9\overline{)0}$	$8\overline{)72}$	$2\overline{)2}$	$5\overline{)40}$	$3\overline{)9}$	$9\overline{)18}$	$1\overline{)6}$	$4\overline{)12}$	$7\overline{)7}$

F

64 Multiplication Facts
For use with Lesson 37

Name _____

Time _____

Multiply.

5 × 6	4 × 3	9 × 8	7 × 5	2 × 9	8 × 4	9 × 3	6 × 9
9 × 4	2 × 5	7 × 6	4 × 8	7 × 9	5 × 4	3 × 2	9 × 7
3 × 7	8 × 5	6 × 2	5 × 5	3 × 5	2 × 4	7 × 7	8 × 9
6 × 4	2 × 8	4 × 4	8 × 2	3 × 9	6 × 6	9 × 9	5 × 3
4 × 6	8 × 8	5 × 7	6 × 3	2 × 2	7 × 4	3 × 8	8 × 6
2 × 6	5 × 9	3 × 3	9 × 2	6 × 7	4 × 5	7 × 2	9 × 6
5 × 2	7 × 8	2 × 3	6 × 8	4 × 7	9 × 5	3 × 6	8 × 7
3 × 4	7 × 3	5 × 8	4 × 2	8 × 3	2 × 7	6 × 5	4 × 9

Saxon Math 6/5—Homeschool

E   90 Division Facts
For use with Lesson 38

Name _____

Time _____

Divide.

20 ÷ 4 =	21 ÷ 7 =	0 ÷ 2 =	27 ÷ 3 =	8 ÷ 1 =	54 ÷ 6 =
15 ÷ 5 =	6 ÷ 3 =	28 ÷ 4 =	18 ÷ 2 =	24 ÷ 6 =	9 ÷ 9 =
56 ÷ 8 =	0 ÷ 6 =	21 ÷ 3 =	1 ÷ 1 =	25 ÷ 5 =	12 ÷ 2 =
5 ÷ 1 =	45 ÷ 9 =	16 ÷ 4 =	30 ÷ 6 =	9 ÷ 3 =	14 ÷ 7 =
0 ÷ 8 =	6 ÷ 2 =	24 ÷ 8 =	10 ÷ 5 =	81 ÷ 9 =	24 ÷ 4 =
16 ÷ 2 =	30 ÷ 5 =	0 ÷ 1 =	28 ÷ 7 =	4 ÷ 4 =	40 ÷ 8 =
3 ÷ 3 =	18 ÷ 6 =	63 ÷ 9 =	40 ÷ 5 =	10 ÷ 2 =	36 ÷ 6 =
32 ÷ 8 =	12 ÷ 4 =	18 ÷ 3 =	35 ÷ 7 =	8 ÷ 8 =	2 ÷ 1 =
45 ÷ 5 =	7 ÷ 7 =	27 ÷ 9 =	9 ÷ 1 =	48 ÷ 6 =	0 ÷ 7 =
4 ÷ 1 =	0 ÷ 9 =	24 ÷ 3 =	32 ÷ 4 =	5 ÷ 5 =	72 ÷ 9 =
56 ÷ 7 =	15 ÷ 3 =	12 ÷ 6 =	8 ÷ 2 =	63 ÷ 7 =	0 ÷ 4 =
14 ÷ 2 =	42 ÷ 6 =	6 ÷ 1 =	16 ÷ 8 =	20 ÷ 5 =	49 ÷ 7 =
36 ÷ 4 =	64 ÷ 8 =	0 ÷ 3 =	54 ÷ 9 =	4 ÷ 2 =	48 ÷ 8 =
18 ÷ 9 =	3 ÷ 1 =	35 ÷ 5 =	8 ÷ 4 =	72 ÷ 8 =	6 ÷ 6 =
0 ÷ 5 =	42 ÷ 7 =	2 ÷ 2 =	36 ÷ 9 =	7 ÷ 1 =	12 ÷ 3 =

F

64 Multiplication Facts
For use with Lesson 39

Name _____

Time _____

Multiply.

5 × 6	4 × 3	9 × 8	7 × 5	2 × 9	8 × 4	9 × 3	6 × 9
9 × 4	2 × 5	7 × 6	4 × 8	7 × 9	5 × 4	3 × 2	9 × 7
3 × 7	8 × 5	6 × 2	5 × 5	3 × 5	2 × 4	7 × 7	8 × 9
6 × 4	2 × 8	4 × 4	8 × 2	3 × 9	6 × 6	9 × 9	5 × 3
4 × 6	8 × 8	5 × 7	6 × 3	2 × 2	7 × 4	3 × 8	8 × 6
2 × 6	5 × 9	3 × 3	9 × 2	6 × 7	4 × 5	7 × 2	9 × 6
5 × 2	7 × 8	2 × 3	6 × 8	4 × 7	9 × 5	3 × 6	8 × 7
3 × 4	7 × 3	5 × 8	4 × 2	8 × 3	2 × 7	6 × 5	4 × 9

Saxon Math 6/5—Homeschool

D 90 Division Facts

For use with Lesson 40

Name _____

Time _____

Divide.

7)21	2)10	6)42	1)3	4)24	3)6	9)54	6)18	4)0	5)30
4)32	8)56	1)0	6)12	3)18	9)72	5)15	2)8	7)42	6)36
6)0	5)10	9)9	2)6	7)63	4)16	8)48	1)2	5)35	3)21
2)18	6)6	3)15	8)40	2)0	5)20	9)27	1)8	4)4	7)35
4)20	9)63	1)4	7)14	3)3	8)24	5)0	6)24	8)8	2)16
5)5	8)64	3)0	4)28	7)49	2)4	9)81	3)12	6)30	1)5
8)32	1)1	9)36	3)27	2)14	5)25	6)48	8)0	7)28	4)36
2)12	5)45	1)7	4)8	7)0	8)16	3)24	9)45	1)9	6)54
7)56	9)0	8)72	2)2	5)40	3)9	9)18	1)6	4)12	7)7

E

90 Division Facts
For use with Test 7

Name _____

Time _____

Divide.

20 ÷ 4 =	21 ÷ 7 =	0 ÷ 2 =	27 ÷ 3 =	8 ÷ 1 =	54 ÷ 6 =
15 ÷ 5 =	6 ÷ 3 =	28 ÷ 4 =	18 ÷ 2 =	24 ÷ 6 =	9 ÷ 9 =
56 ÷ 8 =	0 ÷ 6 =	21 ÷ 3 =	1 ÷ 1 =	25 ÷ 5 =	12 ÷ 2 =
5 ÷ 1 =	45 ÷ 9 =	16 ÷ 4 =	30 ÷ 6 =	9 ÷ 3 =	14 ÷ 7 =
0 ÷ 8 =	6 ÷ 2 =	24 ÷ 8 =	10 ÷ 5 =	81 ÷ 9 =	24 ÷ 4 =
16 ÷ 2 =	30 ÷ 5 =	0 ÷ 1 =	28 ÷ 7 =	4 ÷ 4 =	40 ÷ 8 =
3 ÷ 3 =	18 ÷ 6 =	63 ÷ 9 =	40 ÷ 5 =	10 ÷ 2 =	36 ÷ 6 =
32 ÷ 8 =	12 ÷ 4 =	18 ÷ 3 =	35 ÷ 7 =	8 ÷ 8 =	2 ÷ 1 =
45 ÷ 5 =	7 ÷ 7 =	27 ÷ 9 =	9 ÷ 1 =	48 ÷ 6 =	0 ÷ 7 =
4 ÷ 1 =	0 ÷ 9 =	24 ÷ 3 =	32 ÷ 4 =	5 ÷ 5 =	72 ÷ 9 =
56 ÷ 7 =	15 ÷ 3 =	12 ÷ 6 =	8 ÷ 2 =	63 ÷ 7 =	0 ÷ 4 =
14 ÷ 2 =	42 ÷ 6 =	6 ÷ 1 =	16 ÷ 8 =	20 ÷ 5 =	49 ÷ 7 =
36 ÷ 4 =	64 ÷ 8 =	0 ÷ 3 =	54 ÷ 9 =	4 ÷ 2 =	48 ÷ 8 =
18 ÷ 9 =	3 ÷ 1 =	35 ÷ 5 =	8 ÷ 4 =	72 ÷ 8 =	6 ÷ 6 =
0 ÷ 5 =	42 ÷ 7 =	2 ÷ 2 =	36 ÷ 9 =	7 ÷ 1 =	12 ÷ 3 =

19 Measuring Angles

For use with Investigation 4

Name _____

Use a protractor to measure each angle on this page. Write your measurements inside each angle.

Position the protractor on the angle so that:

1. The protractor is centered on the vertex.
2. One side of the angle passes through one of the 0° marks on the protractor.
3. The other side of the angle passes through the scale.

To decide which scale to read, first note whether you are measuring an acute angle (less than 90°) or an obtuse angle (greater than 90°).

a.

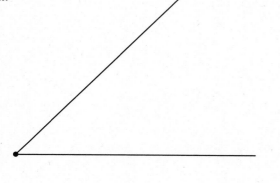

b.

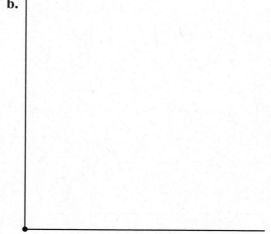

c.   **d.**

F

64 Multiplication Facts
For use with Lesson 41

Name _____

Time _____

Multiply.

5 × 6	4 × 3	9 × 8	7 × 5	2 × 9	8 × 4	9 × 3	6 × 9
9 × 4	2 × 5	7 × 6	4 × 8	7 × 9	5 × 4	3 × 2	9 × 7
3 × 7	8 × 5	6 × 2	5 × 5	3 × 5	2 × 4	7 × 7	8 × 9
6 × 4	2 × 8	4 × 4	8 × 2	3 × 9	6 × 6	9 × 9	5 × 3
4 × 6	8 × 8	5 × 7	6 × 3	2 × 2	7 × 4	3 × 8	8 × 6
2 × 6	5 × 9	3 × 3	9 × 2	6 × 7	4 × 5	7 × 2	9 × 6
5 × 2	7 × 8	2 × 3	6 × 8	4 × 7	9 × 5	3 × 6	8 × 7
3 × 4	7 × 3	5 × 8	4 × 2	8 × 3	2 × 7	6 × 5	4 × 9

E | 90 Division Facts
For use with Lesson 42

Name _____

Time _____

Divide.

20 ÷ 4 =	21 ÷ 7 =	0 ÷ 2 =	27 ÷ 3 =	8 ÷ 1 =	54 ÷ 6 =
15 ÷ 5 =	6 ÷ 3 =	28 ÷ 4 =	18 ÷ 2 =	24 ÷ 6 =	9 ÷ 9 =
56 ÷ 8 =	0 ÷ 6 =	21 ÷ 3 =	1 ÷ 1 =	25 ÷ 5 =	12 ÷ 2 =
5 ÷ 1 =	45 ÷ 9 =	16 ÷ 4 =	30 ÷ 6 =	9 ÷ 3 =	14 ÷ 7 =
0 ÷ 8 =	6 ÷ 2 =	24 ÷ 8 =	10 ÷ 5 =	81 ÷ 9 =	24 ÷ 4 =
16 ÷ 2 =	30 ÷ 5 =	0 ÷ 1 =	28 ÷ 7 =	4 ÷ 4 =	40 ÷ 8 =
3 ÷ 3 =	18 ÷ 6 =	63 ÷ 9 =	40 ÷ 5 =	10 ÷ 2 =	36 ÷ 6 =
32 ÷ 8 =	12 ÷ 4 =	18 ÷ 3 =	35 ÷ 7 =	8 ÷ 8 =	2 ÷ 1 =
45 ÷ 5 =	7 ÷ 7 =	27 ÷ 9 =	9 ÷ 1 =	48 ÷ 6 =	0 ÷ 7 =
4 ÷ 1 =	0 ÷ 9 =	24 ÷ 3 =	32 ÷ 4 =	5 ÷ 5 =	72 ÷ 9 =
56 ÷ 7 =	15 ÷ 3 =	12 ÷ 6 =	8 ÷ 2 =	63 ÷ 7 =	0 ÷ 4 =
14 ÷ 2 =	42 ÷ 6 =	6 ÷ 1 =	16 ÷ 8 =	20 ÷ 5 =	49 ÷ 7 =
36 ÷ 4 =	64 ÷ 8 =	0 ÷ 3 =	54 ÷ 9 =	4 ÷ 2 =	48 ÷ 8 =
18 ÷ 9 =	3 ÷ 1 =	35 ÷ 5 =	8 ÷ 4 =	72 ÷ 8 =	6 ÷ 6 =
0 ÷ 5 =	42 ÷ 7 =	2 ÷ 2 =	36 ÷ 9 =	7 ÷ 1 =	12 ÷ 3 =

D 90 Division Facts
For use with Lesson 43

Name _____

Time _____

Divide.

$7\overline{)21}$	$2\overline{)10}$	$6\overline{)42}$	$1\overline{)3}$	$4\overline{)24}$	$3\overline{)6}$	$9\overline{)54}$	$6\overline{)18}$	$4\overline{)0}$	$5\overline{)30}$
$4\overline{)32}$	$8\overline{)56}$	$1\overline{)0}$	$6\overline{)12}$	$3\overline{)18}$	$9\overline{)72}$	$5\overline{)15}$	$2\overline{)8}$	$7\overline{)42}$	$6\overline{)36}$
$6\overline{)0}$	$5\overline{)10}$	$9\overline{)9}$	$2\overline{)6}$	$7\overline{)63}$	$4\overline{)16}$	$8\overline{)48}$	$1\overline{)2}$	$5\overline{)35}$	$3\overline{)21}$
$2\overline{)18}$	$6\overline{)6}$	$3\overline{)15}$	$8\overline{)40}$	$2\overline{)0}$	$5\overline{)20}$	$9\overline{)27}$	$1\overline{)8}$	$4\overline{)4}$	$7\overline{)35}$
$4\overline{)20}$	$9\overline{)63}$	$1\overline{)4}$	$7\overline{)14}$	$3\overline{)3}$	$8\overline{)24}$	$5\overline{)0}$	$6\overline{)24}$	$8\overline{)8}$	$2\overline{)16}$
$5\overline{)5}$	$8\overline{)64}$	$3\overline{)0}$	$4\overline{)28}$	$7\overline{)49}$	$2\overline{)4}$	$9\overline{)81}$	$3\overline{)12}$	$6\overline{)30}$	$1\overline{)5}$
$8\overline{)32}$	$1\overline{)1}$	$9\overline{)36}$	$3\overline{)27}$	$2\overline{)14}$	$5\overline{)25}$	$6\overline{)48}$	$8\overline{)0}$	$7\overline{)28}$	$4\overline{)36}$
$2\overline{)12}$	$5\overline{)45}$	$1\overline{)7}$	$4\overline{)8}$	$7\overline{)0}$	$8\overline{)16}$	$3\overline{)24}$	$9\overline{)45}$	$1\overline{)9}$	$6\overline{)54}$
$7\overline{)56}$	$9\overline{)0}$	$8\overline{)72}$	$2\overline{)2}$	$5\overline{)40}$	$3\overline{)9}$	$9\overline{)18}$	$1\overline{)6}$	$4\overline{)12}$	$7\overline{)7}$

F	**64 Multiplication Facts**
	For use with Lesson 44

Name _____

Time _____

Multiply.

5 × 6	4 × 3	9 × 8	7 × 5	2 × 9	8 × 4	9 × 3	6 × 9
9 × 4	2 × 5	7 × 6	4 × 8	7 × 9	5 × 4	3 × 2	9 × 7
3 × 7	8 × 5	6 × 2	5 × 5	3 × 5	2 × 4	7 × 7	8 × 9
6 × 4	2 × 8	4 × 4	8 × 2	3 × 9	6 × 6	9 × 9	5 × 3
4 × 6	8 × 8	5 × 7	6 × 3	2 × 2	7 × 4	3 × 8	8 × 6
2 × 6	5 × 9	3 × 3	9 × 2	6 × 7	4 × 5	7 × 2	9 × 6
5 × 2	7 × 8	2 × 3	6 × 8	4 × 7	9 × 5	3 × 6	8 × 7
3 × 4	7 × 3	5 × 8	4 × 2	8 × 3	2 × 7	6 × 5	4 × 9

E | 90 Division Facts
For use with Lesson 45

Name _____

Time _____

Divide.

20 ÷ 4 =	21 ÷ 7 =	0 ÷ 2 =	27 ÷ 3 =	8 ÷ 1 =	54 ÷ 6 =
15 ÷ 5 =	6 ÷ 3 =	28 ÷ 4 =	18 ÷ 2 =	24 ÷ 6 =	9 ÷ 9 =
56 ÷ 8 =	0 ÷ 6 =	21 ÷ 3 =	1 ÷ 1 =	25 ÷ 5 =	12 ÷ 2 =
5 ÷ 1 =	45 ÷ 9 =	16 ÷ 4 =	30 ÷ 6 =	9 ÷ 3 =	14 ÷ 7 =
0 ÷ 8 =	6 ÷ 2 =	24 ÷ 8 =	10 ÷ 5 =	81 ÷ 9 =	24 ÷ 4 =
16 ÷ 2 =	30 ÷ 5 =	0 ÷ 1 =	28 ÷ 7 =	4 ÷ 4 =	40 ÷ 8 =
3 ÷ 3 =	18 ÷ 6 =	63 ÷ 9 =	40 ÷ 5 =	10 ÷ 2 =	36 ÷ 6 =
32 ÷ 8 =	12 ÷ 4 =	18 ÷ 3 =	35 ÷ 7 =	8 ÷ 8 =	2 ÷ 1 =
45 ÷ 5 =	7 ÷ 7 =	27 ÷ 9 =	9 ÷ 1 =	48 ÷ 6 =	0 ÷ 7 =
4 ÷ 1 =	0 ÷ 9 =	24 ÷ 3 =	32 ÷ 4 =	5 ÷ 5 =	72 ÷ 9 =
56 ÷ 7 =	15 ÷ 3 =	12 ÷ 6 =	8 ÷ 2 =	63 ÷ 7 =	0 ÷ 4 =
14 ÷ 2 =	42 ÷ 6 =	6 ÷ 1 =	16 ÷ 8 =	20 ÷ 5 =	49 ÷ 7 =
36 ÷ 4 =	64 ÷ 8 =	0 ÷ 3 =	54 ÷ 9 =	4 ÷ 2 =	48 ÷ 8 =
18 ÷ 9 =	3 ÷ 1 =	35 ÷ 5 =	8 ÷ 4 =	72 ÷ 8 =	6 ÷ 6 =
0 ÷ 5 =	42 ÷ 7 =	2 ÷ 2 =	36 ÷ 9 =	7 ÷ 1 =	12 ÷ 3 =

D · 90 Division Facts
For use with Test 8

Name _____

Time _____

Divide.

7)21	2)10	6)42	1)3	4)24	3)6	9)54	6)18	4)0	5)30
4)32	8)56	1)0	6)12	3)18	9)72	5)15	2)8	7)42	6)36
6)0	5)10	9)9	2)6	7)63	4)16	8)48	1)2	5)35	3)21
2)18	6)6	3)15	8)40	2)0	5)20	9)27	1)8	4)4	7)35
4)20	9)63	1)4	7)14	3)3	8)24	5)0	6)24	8)8	2)16
5)5	8)64	3)0	4)28	7)49	2)4	9)81	3)12	6)30	1)5
8)32	1)1	9)36	3)27	2)14	5)25	6)48	8)0	7)28	4)36
2)12	5)45	1)7	4)8	7)0	8)16	3)24	9)45	1)9	6)54
7)56	9)0	8)72	2)2	5)40	3)9	9)18	1)6	4)12	7)7

Saxon Math 6/5—Homeschool

F

64 Multiplication Facts
For use with Lesson 46

Name _____

Time _____

Multiply.

5 × 6	4 × 3	9 × 8	7 × 5	2 × 9	8 × 4	9 × 3	6 × 9
9 × 4	2 × 5	7 × 6	4 × 8	7 × 9	5 × 4	3 × 2	9 × 7
3 × 7	8 × 5	6 × 2	5 × 5	3 × 5	2 × 4	7 × 7	8 × 9
6 × 4	2 × 8	4 × 4	8 × 2	3 × 9	6 × 6	9 × 9	5 × 3
4 × 6	8 × 8	5 × 7	6 × 3	2 × 2	7 × 4	3 × 8	8 × 6
2 × 6	5 × 9	3 × 3	9 × 2	6 × 7	4 × 5	7 × 2	9 × 6
5 × 2	7 × 8	2 × 3	6 × 8	4 × 7	9 × 5	3 × 6	8 × 7
3 × 4	7 × 3	5 × 8	4 × 2	8 × 3	2 × 7	6 × 5	4 × 9

F

64 Multiplication Facts
For use with Lesson 47

Name _____

Time _____

Multiply.

5 × 6	4 × 3	9 × 8	7 × 5	2 × 9	8 × 4	9 × 3	6 × 9
9 × 4	2 × 5	7 × 6	4 × 8	7 × 9	5 × 4	3 × 2	9 × 7
3 × 7	8 × 5	6 × 2	5 × 5	3 × 5	2 × 4	7 × 7	8 × 9
6 × 4	2 × 8	4 × 4	8 × 2	3 × 9	6 × 6	9 × 9	5 × 3
4 × 6	8 × 8	5 × 7	6 × 3	2 × 2	7 × 4	3 × 8	8 × 6
2 × 6	5 × 9	3 × 3	9 × 2	6 × 7	4 × 5	7 × 2	9 × 6
5 × 2	7 × 8	2 × 3	6 × 8	4 × 7	9 × 5	3 × 6	8 × 7
3 × 4	7 × 3	5 × 8	4 × 2	8 × 3	2 × 7	6 × 5	4 × 9

<table>
<tr><td>**G**</td><td>**48 Uneven Divisions**
For use with Lesson 48</td><td>Name _____

Time _____</td></tr>
</table>

Divide. Write each answer with a remainder.

$4\overline{)15}$	$9\overline{)14}$	$7\overline{)45}$	$3\overline{)16}$	$6\overline{)38}$	$2\overline{)7}$
$8\overline{)50}$	$5\overline{)28}$	$4\overline{)21}$	$6\overline{)15}$	$7\overline{)11}$	$8\overline{)20}$
$3\overline{)20}$	$7\overline{)32}$	$8\overline{)30}$	$2\overline{)15}$	$5\overline{)43}$	$6\overline{)35}$
$9\overline{)62}$	$4\overline{)10}$	$6\overline{)27}$	$9\overline{)21}$	$4\overline{)19}$	$3\overline{)25}$
$6\overline{)56}$	$2\overline{)17}$	$3\overline{)10}$	$5\overline{)8}$	$9\overline{)40}$	$7\overline{)30}$
$2\overline{)5}$	$8\overline{)25}$	$5\overline{)17}$	$7\overline{)17}$	$3\overline{)8}$	$4\overline{)9}$
$7\overline{)20}$	$6\overline{)10}$	$2\overline{)9}$	$4\overline{)30}$	$8\overline{)15}$	$9\overline{)29}$
$5\overline{)32}$	$3\overline{)14}$	$9\overline{)50}$	$8\overline{)65}$	$2\overline{)11}$	$5\overline{)19}$

G

48 Uneven Divisions
For use with Lesson 49

Name _____

Time _____

Divide. Write each answer with a remainder.

$4\overline{)15}$	$9\overline{)14}$	$7\overline{)45}$	$3\overline{)16}$	$6\overline{)38}$	$2\overline{)7}$
$8\overline{)50}$	$5\overline{)28}$	$4\overline{)21}$	$6\overline{)15}$	$7\overline{)11}$	$8\overline{)20}$
$3\overline{)20}$	$7\overline{)32}$	$8\overline{)30}$	$2\overline{)15}$	$5\overline{)43}$	$6\overline{)35}$
$9\overline{)62}$	$4\overline{)10}$	$6\overline{)27}$	$9\overline{)21}$	$4\overline{)19}$	$3\overline{)25}$
$6\overline{)56}$	$2\overline{)17}$	$3\overline{)10}$	$5\overline{)8}$	$9\overline{)40}$	$7\overline{)30}$
$2\overline{)5}$	$8\overline{)25}$	$5\overline{)17}$	$7\overline{)17}$	$3\overline{)8}$	$4\overline{)9}$
$7\overline{)20}$	$6\overline{)10}$	$2\overline{)9}$	$4\overline{)30}$	$8\overline{)15}$	$9\overline{)29}$
$5\overline{)32}$	$3\overline{)14}$	$9\overline{)50}$	$8\overline{)65}$	$2\overline{)11}$	$5\overline{)19}$

F

64 Multiplication Facts
For use with Lesson 50

Name _____

Time _____

Multiply.

5 × 6	4 × 3	9 × 8	7 × 5	2 × 9	8 × 4	9 × 3	6 × 9
9 × 4	2 × 5	7 × 6	4 × 8	7 × 9	5 × 4	3 × 2	9 × 7
3 × 7	8 × 5	6 × 2	5 × 5	3 × 5	2 × 4	7 × 7	8 × 9
6 × 4	2 × 8	4 × 4	8 × 2	3 × 9	6 × 6	9 × 9	5 × 3
4 × 6	8 × 8	5 × 7	6 × 3	2 × 2	7 × 4	3 × 8	8 × 6
2 × 6	5 × 9	3 × 3	9 × 2	6 × 7	4 × 5	7 × 2	9 × 6
5 × 2	7 × 8	2 × 3	6 × 8	4 × 7	9 × 5	3 × 6	8 × 7
3 × 4	7 × 3	5 × 8	4 × 2	8 × 3	2 × 7	6 × 5	4 × 9

F | 64 Multiplication Facts

For use with Test 9

Name _____

Time _____

Multiply.

5 × 6	4 × 3	9 × 8	7 × 5	2 × 9	8 × 4	9 × 3	6 × 9
9 × 4	2 × 5	7 × 6	4 × 8	7 × 9	5 × 4	3 × 2	9 × 7
3 × 7	8 × 5	6 × 2	5 × 5	3 × 5	2 × 4	7 × 7	8 × 9
6 × 4	2 × 8	4 × 4	8 × 2	3 × 9	6 × 6	9 × 9	5 × 3
4 × 6	8 × 8	5 × 7	6 × 3	2 × 2	7 × 4	3 × 8	8 × 6
2 × 6	5 × 9	3 × 3	9 × 2	6 × 7	4 × 5	7 × 2	9 × 6
5 × 2	7 × 8	2 × 3	6 × 8	4 × 7	9 × 5	3 × 6	8 × 7
3 × 4	7 × 3	5 × 8	4 × 2	8 × 3	2 × 7	6 × 5	4 × 9

Saxon Math 6/5—Homeschool

G | 48 Uneven Divisions

For use with Lesson 51

Name _____

Time _____

Divide. Write each answer with a remainder.

4)15	9)14	7)45	3)16	6)38	2)7
8)50	5)28	4)21	6)15	7)11	8)20
3)20	7)32	8)30	2)15	5)43	6)35
9)62	4)10	6)27	9)21	4)19	3)25
6)56	2)17	3)10	5)8	9)40	7)30
2)5	8)25	5)17	7)17	3)8	4)9
7)20	6)10	2)9	4)30	8)15	9)29
5)32	3)14	9)50	8)65	2)11	5)19

F

64 Multiplication Facts
For use with Lesson 52

Name _____

Time _____

Multiply.

5 × 6	4 × 3	9 × 8	7 × 5	2 × 9	8 × 4	9 × 3	6 × 9
9 × 4	2 × 5	7 × 6	4 × 8	7 × 9	5 × 4	3 × 2	9 × 7
3 × 7	8 × 5	6 × 2	5 × 5	3 × 5	2 × 4	7 × 7	8 × 9
6 × 4	2 × 8	4 × 4	8 × 2	3 × 9	6 × 6	9 × 9	5 × 3
4 × 6	8 × 8	5 × 7	6 × 3	2 × 2	7 × 4	3 × 8	8 × 6
2 × 6	5 × 9	3 × 3	9 × 2	6 × 7	4 × 5	7 × 2	9 × 6
5 × 2	7 × 8	2 × 3	6 × 8	4 × 7	9 × 5	3 × 6	8 × 7
3 × 4	7 × 3	5 × 8	4 × 2	8 × 3	2 × 7	6 × 5	4 × 9

F | 64 Multiplication Facts

For use with Lesson 53

Name _____

Time _____

Multiply.

5 × 6	4 × 3	9 × 8	7 × 5	2 × 9	8 × 4	9 × 3	6 × 9
9 × 4	2 × 5	7 × 6	4 × 8	7 × 9	5 × 4	3 × 2	9 × 7
3 × 7	8 × 5	6 × 2	5 × 5	3 × 5	2 × 4	7 × 7	8 × 9
6 × 4	2 × 8	4 × 4	8 × 2	3 × 9	6 × 6	9 × 9	5 × 3
4 × 6	8 × 8	5 × 7	6 × 3	2 × 2	7 × 4	3 × 8	8 × 6
2 × 6	5 × 9	3 × 3	9 × 2	6 × 7	4 × 5	7 × 2	9 × 6
5 × 2	7 × 8	2 × 3	6 × 8	4 × 7	9 × 5	3 × 6	8 × 7
3 × 4	7 × 3	5 × 8	4 × 2	8 × 3	2 × 7	6 × 5	4 × 9

20 Measuring Circles

For use with Lesson 53

Name _____

To measure the radius and the diameter of a circle or circular object, first locate the center of the circle. The **radius** is the distance from the center to the circle. The **diameter** is the distance across the circle through its center.

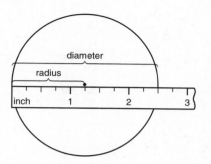

The **circumference** is the distance around the circle. To find the circumference of a circular object wrap a cloth tape measure around the object. If a tape measure is not available, wrap masking tape or string around the object. Then unwrap the string or tape and measure the part that wrapped once around the object.

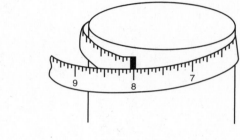

Another way to measure a circumference is to make a mark on the edge of the object and roll the object once around. The distance between the two points where the mark touched down is the circumference.

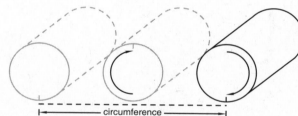

Measure the diameter, the radius, and the circumference of several circular objects.

Object	Diameter	Radius	Circumference

G 48 Uneven Divisions
For use with Lesson 54

Name _____

Time _____

Divide. Write each answer with a remainder.

$4\overline{)15}$	$9\overline{)14}$	$7\overline{)45}$	$3\overline{)16}$	$6\overline{)38}$	$2\overline{)7}$
$8\overline{)50}$	$5\overline{)28}$	$4\overline{)21}$	$6\overline{)15}$	$7\overline{)11}$	$8\overline{)20}$
$3\overline{)20}$	$7\overline{)32}$	$8\overline{)30}$	$2\overline{)15}$	$5\overline{)43}$	$6\overline{)35}$
$9\overline{)62}$	$4\overline{)10}$	$6\overline{)27}$	$9\overline{)21}$	$4\overline{)19}$	$3\overline{)25}$
$6\overline{)56}$	$2\overline{)17}$	$3\overline{)10}$	$5\overline{)8}$	$9\overline{)40}$	$7\overline{)30}$
$2\overline{)5}$	$8\overline{)25}$	$5\overline{)17}$	$7\overline{)17}$	$3\overline{)8}$	$4\overline{)9}$
$7\overline{)20}$	$6\overline{)10}$	$2\overline{)9}$	$4\overline{)30}$	$8\overline{)15}$	$9\overline{)29}$
$5\overline{)32}$	$3\overline{)14}$	$9\overline{)50}$	$8\overline{)65}$	$2\overline{)11}$	$5\overline{)19}$

| F |

64 Multiplication Facts
For use with Lesson 55

Name _____

Time _____

Multiply.

5 × 6	4 × 3	9 × 8	7 × 5	2 × 9	8 × 4	9 × 3	6 × 9
9 × 4	2 × 5	7 × 6	4 × 8	7 × 9	5 × 4	3 × 2	9 × 7
3 × 7	8 × 5	6 × 2	5 × 5	3 × 5	2 × 4	7 × 7	8 × 9
6 × 4	2 × 8	4 × 4	8 × 2	3 × 9	6 × 6	9 × 9	5 × 3
4 × 6	8 × 8	5 × 7	6 × 3	2 × 2	7 × 4	3 × 8	8 × 6
2 × 6	5 × 9	3 × 3	9 × 2	6 × 7	4 × 5	7 × 2	9 × 6
5 × 2	7 × 8	2 × 3	6 × 8	4 × 7	9 × 5	3 × 6	8 × 7
3 × 4	7 × 3	5 × 8	4 × 2	8 × 3	2 × 7	6 × 5	4 × 9

F | 64 Multiplication Facts
For use with Test 10

Name _____

Time _____

Multiply.

5 × 6	4 × 3	9 × 8	7 × 5	2 × 9	8 × 4	9 × 3	6 × 9
9 × 4	2 × 5	7 × 6	4 × 8	7 × 9	5 × 4	3 × 2	9 × 7
3 × 7	8 × 5	6 × 2	5 × 5	3 × 5	2 × 4	7 × 7	8 × 9
6 × 4	2 × 8	4 × 4	8 × 2	3 × 9	6 × 6	9 × 9	5 × 3
4 × 6	8 × 8	5 × 7	6 × 3	2 × 2	7 × 4	3 × 8	8 × 6
2 × 6	5 × 9	3 × 3	9 × 2	6 × 7	4 × 5	7 × 2	9 × 6
5 × 2	7 × 8	2 × 3	6 × 8	4 × 7	9 × 5	3 × 6	8 × 7
3 × 4	7 × 3	5 × 8	4 × 2	8 × 3	2 × 7	6 × 5	4 × 9

G	**48 Uneven Divisions**

For use with Lesson 56

Name _____

Time _____

Divide. Write each answer with a remainder.

$4\overline{)15}$	$9\overline{)14}$	$7\overline{)45}$	$3\overline{)16}$	$6\overline{)38}$	$2\overline{)7}$
$8\overline{)50}$	$5\overline{)28}$	$4\overline{)21}$	$6\overline{)15}$	$7\overline{)11}$	$8\overline{)20}$
$3\overline{)20}$	$7\overline{)32}$	$8\overline{)30}$	$2\overline{)15}$	$5\overline{)43}$	$6\overline{)35}$
$9\overline{)62}$	$4\overline{)10}$	$6\overline{)27}$	$9\overline{)21}$	$4\overline{)19}$	$3\overline{)25}$
$6\overline{)56}$	$2\overline{)17}$	$3\overline{)10}$	$5\overline{)8}$	$9\overline{)40}$	$7\overline{)30}$
$2\overline{)5}$	$8\overline{)25}$	$5\overline{)17}$	$7\overline{)17}$	$3\overline{)8}$	$4\overline{)9}$
$7\overline{)20}$	$6\overline{)10}$	$2\overline{)9}$	$4\overline{)30}$	$8\overline{)15}$	$9\overline{)29}$
$5\overline{)32}$	$3\overline{)14}$	$9\overline{)50}$	$8\overline{)65}$	$2\overline{)11}$	$5\overline{)19}$

F

64 Multiplication Facts

For use with Lesson 57

Name _____

Time _____

Multiply.

5 × 6	4 × 3	9 × 8	7 × 5	2 × 9	8 × 4	9 × 3	6 × 9
9 × 4	2 × 5	7 × 6	4 × 8	7 × 9	5 × 4	3 × 2	9 × 7
3 × 7	8 × 5	6 × 2	5 × 5	3 × 5	2 × 4	7 × 7	8 × 9
6 × 4	2 × 8	4 × 4	8 × 2	3 × 9	6 × 6	9 × 9	5 × 3
4 × 6	8 × 8	5 × 7	6 × 3	2 × 2	7 × 4	3 × 8	8 × 6
2 × 6	5 × 9	3 × 3	9 × 2	6 × 7	4 × 5	7 × 2	9 × 6
5 × 2	7 × 8	2 × 3	6 × 8	4 × 7	9 × 5	3 × 6	8 × 7
3 × 4	7 × 3	5 × 8	4 × 2	8 × 3	2 × 7	6 × 5	4 × 9

G	**48 Uneven Divisions**
	For use with Lesson 58

Name _____

Time _____

Divide. Write each answer with a remainder.

$4\overline{)15}$	$9\overline{)14}$	$7\overline{)45}$	$3\overline{)16}$	$6\overline{)38}$	$2\overline{)7}$
$8\overline{)50}$	$5\overline{)28}$	$4\overline{)21}$	$6\overline{)15}$	$7\overline{)11}$	$8\overline{)20}$
$3\overline{)20}$	$7\overline{)32}$	$8\overline{)30}$	$2\overline{)15}$	$5\overline{)43}$	$6\overline{)35}$
$9\overline{)62}$	$4\overline{)10}$	$6\overline{)27}$	$9\overline{)21}$	$4\overline{)19}$	$3\overline{)25}$
$6\overline{)56}$	$2\overline{)17}$	$3\overline{)10}$	$5\overline{)8}$	$9\overline{)40}$	$7\overline{)30}$
$2\overline{)5}$	$8\overline{)25}$	$5\overline{)17}$	$7\overline{)17}$	$3\overline{)8}$	$4\overline{)9}$
$7\overline{)20}$	$6\overline{)10}$	$2\overline{)9}$	$4\overline{)30}$	$8\overline{)15}$	$9\overline{)29}$
$5\overline{)32}$	$3\overline{)14}$	$9\overline{)50}$	$8\overline{)65}$	$2\overline{)11}$	$5\overline{)19}$

F

64 Multiplication Facts
For use with Lesson 59

Name _____

Time _____

Multiply.

5 × 6	4 × 3	9 × 8	7 × 5	2 × 9	8 × 4	9 × 3	6 × 9
9 × 4	2 × 5	7 × 6	4 × 8	7 × 9	5 × 4	3 × 2	9 × 7
3 × 7	8 × 5	6 × 2	5 × 5	3 × 5	2 × 4	7 × 7	8 × 9
6 × 4	2 × 8	4 × 4	8 × 2	3 × 9	6 × 6	9 × 9	5 × 3
4 × 6	8 × 8	5 × 7	6 × 3	2 × 2	7 × 4	3 × 8	8 × 6
2 × 6	5 × 9	3 × 3	9 × 2	6 × 7	4 × 5	7 × 2	9 × 6
5 × 2	7 × 8	2 × 3	6 × 8	4 × 7	9 × 5	3 × 6	8 × 7
3 × 4	7 × 3	5 × 8	4 × 2	8 × 3	2 × 7	6 × 5	4 × 9

G	**48 Uneven Divisions**
	For use with Lesson 60

Name _____

Time _____

Divide. Write each answer with a remainder.

$4\overline{)15}$	$9\overline{)14}$	$7\overline{)45}$	$3\overline{)16}$	$6\overline{)38}$	$2\overline{)7}$
$8\overline{)50}$	$5\overline{)28}$	$4\overline{)21}$	$6\overline{)15}$	$7\overline{)11}$	$8\overline{)20}$
$3\overline{)20}$	$7\overline{)32}$	$8\overline{)30}$	$2\overline{)15}$	$5\overline{)43}$	$6\overline{)35}$
$9\overline{)62}$	$4\overline{)10}$	$6\overline{)27}$	$9\overline{)21}$	$4\overline{)19}$	$3\overline{)25}$
$6\overline{)56}$	$2\overline{)17}$	$3\overline{)10}$	$5\overline{)8}$	$9\overline{)40}$	$7\overline{)30}$
$2\overline{)5}$	$8\overline{)25}$	$5\overline{)17}$	$7\overline{)17}$	$3\overline{)8}$	$4\overline{)9}$
$7\overline{)20}$	$6\overline{)10}$	$2\overline{)9}$	$4\overline{)30}$	$8\overline{)15}$	$9\overline{)29}$
$5\overline{)32}$	$3\overline{)14}$	$9\overline{)50}$	$8\overline{)65}$	$2\overline{)11}$	$5\overline{)19}$

G | 48 Uneven Divisions

For use with Test 11

Name _____

Time _____

Divide. Write each answer with a remainder.

4)15̄	9)14̄	7)45̄	3)16̄	6)38̄	2)7̄
8)50̄	5)28̄	4)21̄	6)15̄	7)11̄	8)20̄
3)20̄	7)32̄	8)30̄	2)15̄	5)43̄	6)35̄
9)62̄	4)10̄	6)27̄	9)21̄	4)19̄	3)25̄
6)56̄	2)17̄	3)10̄	5)8̄	9)40̄	7)30̄
2)5̄	8)25̄	5)17̄	7)17̄	3)8̄	4)9̄
7)20̄	6)10̄	2)9̄	4)30̄	8)15̄	9)29̄
5)32̄	3)14̄	9)50̄	8)65̄	2)11̄	5)19̄

21 | Frequency Tables

For use with Investigation 6

Name _____

Probability Experiment 1

Upturned Number	Tally	Frequency
1		
2		
3		
4		
5		
6		

Probability Experiment 2

Letter	Tally	Relative Frequency
C		
A		
T		

F

64 Multiplication Facts
For use with Lesson 61

Name _____

Time _____

Multiply.

5 × 6	4 × 3	9 × 8	7 × 5	2 × 9	8 × 4	9 × 3	6 × 9
9 × 4	2 × 5	7 × 6	4 × 8	7 × 9	5 × 4	3 × 2	9 × 7
3 × 7	8 × 5	6 × 2	5 × 5	3 × 5	2 × 4	7 × 7	8 × 9
6 × 4	2 × 8	4 × 4	8 × 2	3 × 9	6 × 6	9 × 9	5 × 3
4 × 6	8 × 8	5 × 7	6 × 3	2 × 2	7 × 4	3 × 8	8 × 6
2 × 6	5 × 9	3 × 3	9 × 2	6 × 7	4 × 5	7 × 2	9 × 6
5 × 2	7 × 8	2 × 3	6 × 8	4 × 7	9 × 5	3 × 6	8 × 7
3 × 4	7 × 3	5 × 8	4 × 2	8 × 3	2 × 7	6 × 5	4 × 9

G

48 Uneven Divisions

For use with Lesson 62

Name _____

Time _____

Divide. Write each answer with a remainder.

$4\overline{)15}$	$9\overline{)14}$	$7\overline{)45}$	$3\overline{)16}$	$6\overline{)38}$	$2\overline{)7}$
$8\overline{)50}$	$5\overline{)28}$	$4\overline{)21}$	$6\overline{)15}$	$7\overline{)11}$	$8\overline{)20}$
$3\overline{)20}$	$7\overline{)32}$	$8\overline{)30}$	$2\overline{)15}$	$5\overline{)43}$	$6\overline{)35}$
$9\overline{)62}$	$4\overline{)10}$	$6\overline{)27}$	$9\overline{)21}$	$4\overline{)19}$	$3\overline{)25}$
$6\overline{)56}$	$2\overline{)17}$	$3\overline{)10}$	$5\overline{)8}$	$9\overline{)40}$	$7\overline{)30}$
$2\overline{)5}$	$8\overline{)25}$	$5\overline{)17}$	$7\overline{)17}$	$3\overline{)8}$	$4\overline{)9}$
$7\overline{)20}$	$6\overline{)10}$	$2\overline{)9}$	$4\overline{)30}$	$8\overline{)15}$	$9\overline{)29}$
$5\overline{)32}$	$3\overline{)14}$	$9\overline{)50}$	$8\overline{)65}$	$2\overline{)11}$	$5\overline{)19}$

F	**64 Multiplication Facts**

For use with Lesson 63

Name _____

Time _____

Multiply.

5 × 6	4 × 3	9 × 8	7 × 5	2 × 9	8 × 4	9 × 3	6 × 9
9 × 4	2 × 5	7 × 6	4 × 8	7 × 9	5 × 4	3 × 2	9 × 7
3 × 7	8 × 5	6 × 2	5 × 5	3 × 5	2 × 4	7 × 7	8 × 9
6 × 4	2 × 8	4 × 4	8 × 2	3 × 9	6 × 6	9 × 9	5 × 3
4 × 6	8 × 8	5 × 7	6 × 3	2 × 2	7 × 4	3 × 8	8 × 6
2 × 6	5 × 9	3 × 3	9 × 2	6 × 7	4 × 5	7 × 2	9 × 6
5 × 2	7 × 8	2 × 3	6 × 8	4 × 7	9 × 5	3 × 6	8 × 7
3 × 4	7 × 3	5 × 8	4 × 2	8 × 3	2 × 7	6 × 5	4 × 9

F

64 Multiplication Facts
For use with Lesson 64

Name _____

Time _____

Multiply.

5 × 6	4 × 3	9 × 8	7 × 5	2 × 9	8 × 4	9 × 3	6 × 9
9 × 4	2 × 5	7 × 6	4 × 8	7 × 9	5 × 4	3 × 2	9 × 7
3 × 7	8 × 5	6 × 2	5 × 5	3 × 5	2 × 4	7 × 7	8 × 9
6 × 4	2 × 8	4 × 4	8 × 2	3 × 9	6 × 6	9 × 9	5 × 3
4 × 6	8 × 8	5 × 7	6 × 3	2 × 2	7 × 4	3 × 8	8 × 6
2 × 6	5 × 9	3 × 3	9 × 2	6 × 7	4 × 5	7 × 2	9 × 6
5 × 2	7 × 8	2 × 3	6 × 8	4 × 7	9 × 5	3 × 6	8 × 7
3 × 4	7 × 3	5 × 8	4 × 2	8 × 3	2 × 7	6 × 5	4 × 9

C			

100 Multiplication Facts
For use with Lesson 65

Name _____

Time _____

Multiply.

9 × 9	3 × 5	8 × 5	2 × 6	4 × 7	0 × 3	7 × 2	1 × 5	7 × 8	4 × 0
3 × 4	5 × 9	0 × 2	7 × 3	4 × 1	2 × 7	6 × 3	5 × 4	1 × 0	9 × 2
1 × 1	9 × 0	2 × 8	6 × 4	0 × 7	8 × 1	3 × 3	4 × 8	9 × 3	2 × 0
4 × 9	7 × 0	1 × 2	8 × 4	6 × 5	2 × 9	9 × 4	0 × 1	7 × 4	5 × 8
0 × 8	4 × 2	9 × 8	3 × 6	5 × 5	1 × 6	5 × 0	6 × 6	2 × 1	7 × 9
9 × 1	2 × 2	5 × 1	4 × 3	0 × 0	8 × 9	3 × 7	9 × 7	1 × 7	6 × 0
5 × 6	7 × 5	3 × 0	8 × 8	1 × 3	8 × 3	5 × 2	0 × 4	9 × 5	6 × 7
2 × 3	8 × 6	0 × 5	6 × 1	3 × 8	7 × 6	1 × 8	9 × 6	4 × 4	5 × 3
7 × 7	1 × 4	6 × 2	4 × 5	2 × 4	8 × 0	3 × 1	6 × 8	0 × 9	8 × 7
3 × 2	4 × 6	1 × 9	5 × 7	8 × 2	0 × 6	7 × 1	2 × 5	6 × 9	3 × 9

22 | Metric Unit Strips

For use with Lesson 65

Below are decimeter, centimeter, and millimeter strips. Cut out the strips and glue or tape them to Activity Sheet 23 to make the five lengths shown. You will need to cut the centimeter strips and millimeter strips into smaller pieces to complete the activity.

1 decimeter (10 cm, 100 mm)
1 decimeter (10 cm, 100 mm)
1 decimeter (10 cm, 100 mm)
1 decimeter (10 cm, 100 mm)
1 decimeter (10 cm, 100 mm)

1 cm	1 cm	1 cm	1 cm	1 cm	1 cm	1 cm	1 cm	1 cm	1 cm

1 cm	1 cm	1 cm	1 cm	1 cm	1 cm	1 cm	1 cm	1 cm	1 cm

1 cm	1 cm	1 cm	1 cm	1 cm	1 cm	1 cm	1 cm	1 cm	1 cm

10 mm	10 mm	10 mm	10 mm	10 mm	10 mm	10 mm	10 mm	10 mm	10 mm

23

Decimal Parts of a Meter   Name _____

For use with Lesson 65

Paste or tape the metric unit strips from Activity Sheet 22 onto this page to make each length. The first length is marked as an example to show which strips to use and where to paste them. Before making each length, decide which metric units to use and how many of each unit are needed. You will need to cut the centimeter and millimeter strips into smaller pieces to complete this activity.

0.15 m	1 decimeter		1 cm	1 cm	1 cm	1 cm	1 cm
0.123 m							
0.1 m							
0.075 m							
0.105 m							

Refer to the lengths you made to complete these problems.

1. Write the five lengths above in order from shortest to longest.

_____ , _____ , _____ , _____ , _____

2. Compare:

   a. 0.15 m ◯ 0.123 m   b. 0.075 m ◯ 0.1 m

   c. 0.15 m ◯ 0.075 m   d. 0.105 m ◯ 0.15 m

3. Convert:

   a. 0.15 meters is how many centimeters? _____

   b. 0.123 meters is how many millimeters? _____

4. Round:

   a. Is 0.123 meters closer in length to 0.1 meters or 0.15 meters? _____

   b. Is 0.123 meters closer in length to 0.1 meters or 0.2 meters? _____

5. Add by counting the parts:

   a. 0.15 m + 0.123 m = _____   b. 0.15 m + 0.1 m = _____

FACTS PRACTICE TEST

F

64 Multiplication Facts
For use with Test 12

Name _____

Time _____

Multiply.

5 × 6	4 × 3	9 × 8	7 × 5	2 × 9	8 × 4	9 × 3	6 × 9
9 × 4	2 × 5	7 × 6	4 × 8	7 × 9	5 × 4	3 × 2	9 × 7
3 × 7	8 × 5	6 × 2	5 × 5	3 × 5	2 × 4	7 × 7	8 × 9
6 × 4	2 × 8	4 × 4	8 × 2	3 × 9	6 × 6	9 × 9	5 × 3
4 × 6	8 × 8	5 × 7	6 × 3	2 × 2	7 × 4	3 × 8	8 × 6
2 × 6	5 × 9	3 × 3	9 × 2	6 × 7	4 × 5	7 × 2	9 × 6
5 × 2	7 × 8	2 × 3	6 × 8	4 × 7	9 × 5	3 × 6	8 × 7
3 × 4	7 × 3	5 × 8	4 × 2	8 × 3	2 × 7	6 × 5	4 × 9

F

64 Multiplication Facts
For use with Lesson 66

Name _____

Time _____

Multiply.

5 × 6	4 × 3	9 × 8	7 × 5	2 × 9	8 × 4	9 × 3	6 × 9
9 × 4	2 × 5	7 × 6	4 × 8	7 × 9	5 × 4	3 × 2	9 × 7
3 × 7	8 × 5	6 × 2	5 × 5	3 × 5	2 × 4	7 × 7	8 × 9
6 × 4	2 × 8	4 × 4	8 × 2	3 × 9	6 × 6	9 × 9	5 × 3
4 × 6	8 × 8	5 × 7	6 × 3	2 × 2	7 × 4	3 × 8	8 × 6
2 × 6	5 × 9	3 × 3	9 × 2	6 × 7	4 × 5	7 × 2	9 × 6
5 × 2	7 × 8	2 × 3	6 × 8	4 × 7	9 × 5	3 × 6	8 × 7
3 × 4	7 × 3	5 × 8	4 × 2	8 × 3	2 × 7	6 × 5	4 × 9

G	**48 Uneven Divisions**	Name _____
	For use with Lesson 67	Time _____

Divide. Write each answer with a remainder.

$4\overline{)15}$	$9\overline{)14}$	$7\overline{)45}$	$3\overline{)16}$	$6\overline{)38}$	$2\overline{)7}$
$8\overline{)50}$	$5\overline{)28}$	$4\overline{)21}$	$6\overline{)15}$	$7\overline{)11}$	$8\overline{)20}$
$3\overline{)20}$	$7\overline{)32}$	$8\overline{)30}$	$2\overline{)15}$	$5\overline{)43}$	$6\overline{)35}$
$9\overline{)62}$	$4\overline{)10}$	$6\overline{)27}$	$9\overline{)21}$	$4\overline{)19}$	$3\overline{)25}$
$6\overline{)56}$	$2\overline{)17}$	$3\overline{)10}$	$5\overline{)8}$	$9\overline{)40}$	$7\overline{)30}$
$2\overline{)5}$	$8\overline{)25}$	$5\overline{)17}$	$7\overline{)17}$	$3\overline{)8}$	$4\overline{)9}$
$7\overline{)20}$	$6\overline{)10}$	$2\overline{)9}$	$4\overline{)30}$	$8\overline{)15}$	$9\overline{)29}$
$5\overline{)32}$	$3\overline{)14}$	$9\overline{)50}$	$8\overline{)65}$	$2\overline{)11}$	$5\overline{)19}$

24 | Comparing Decimal Numbers

For use with Lesson 67

Name _____

Shade the squares to represent each decimal number. Then compare the decimal numbers by comparing the shaded part of each square. The first problem is marked as an example.

   0.4 ⓥ 0.33

1.   0.1 ◯ 0.01

2.   0.12 ◯ 0.21

3.   0.3 ◯ 0.30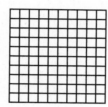

4.   0.5 ◯ 0.05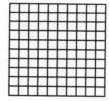

5.   0.6 ◯ 0.67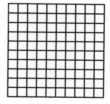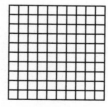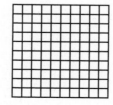

D

90 Division Facts

For use with Lesson 68

Name _____

Time _____

Divide.

7)21	2)10	6)42	1)3	4)24	3)6	9)54	6)18	4)0	5)30
4)32	8)56	1)0	6)12	3)18	9)72	5)15	2)8	7)42	6)36
6)0	5)10	9)9	2)6	7)63	4)16	8)48	1)2	5)35	3)21
2)18	6)6	3)15	8)40	2)0	5)20	9)27	1)8	4)4	7)35
4)20	9)63	1)4	7)14	3)3	8)24	5)0	6)24	8)8	2)16
5)5	8)64	3)0	4)28	7)49	2)4	9)81	3)12	6)30	1)5
8)32	1)1	9)36	3)27	2)14	5)25	6)48	8)0	7)28	4)36
2)12	5)45	1)7	4)8	7)0	8)16	3)24	9)45	1)9	6)54
7)56	9)0	8)72	2)2	5)40	3)9	9)18	1)6	4)12	7)7

F

64 Multiplication Facts
For use with Lesson 69

Name _____

Time _____

Multiply.

5 × 6	4 × 3	9 × 8	7 × 5	2 × 9	8 × 4	9 × 3	6 × 9
9 × 4	2 × 5	7 × 6	4 × 8	7 × 9	5 × 4	3 × 2	9 × 7
3 × 7	8 × 5	6 × 2	5 × 5	3 × 5	2 × 4	7 × 7	8 × 9
6 × 4	2 × 8	4 × 4	8 × 2	3 × 9	6 × 6	9 × 9	5 × 3
4 × 6	8 × 8	5 × 7	6 × 3	2 × 2	7 × 4	3 × 8	8 × 6
2 × 6	5 × 9	3 × 3	9 × 2	6 × 7	4 × 5	7 × 2	9 × 6
5 × 2	7 × 8	2 × 3	6 × 8	4 × 7	9 × 5	3 × 6	8 × 7
3 × 4	7 × 3	5 × 8	4 × 2	8 × 3	2 × 7	6 × 5	4 × 9

G

48 Uneven Divisions

For use with Lesson 70

Name _____

Time _____

Divide. Write each answer with a remainder.

$4\overline{)15}$	$9\overline{)14}$	$7\overline{)45}$	$3\overline{)16}$	$6\overline{)38}$	$2\overline{)7}$
$8\overline{)50}$	$5\overline{)28}$	$4\overline{)21}$	$6\overline{)15}$	$7\overline{)11}$	$8\overline{)20}$
$3\overline{)20}$	$7\overline{)32}$	$8\overline{)30}$	$2\overline{)15}$	$5\overline{)43}$	$6\overline{)35}$
$9\overline{)62}$	$4\overline{)10}$	$6\overline{)27}$	$9\overline{)21}$	$4\overline{)19}$	$3\overline{)25}$
$6\overline{)56}$	$2\overline{)17}$	$3\overline{)10}$	$5\overline{)8}$	$9\overline{)40}$	$7\overline{)30}$
$2\overline{)5}$	$8\overline{)25}$	$5\overline{)17}$	$7\overline{)17}$	$3\overline{)8}$	$4\overline{)9}$
$7\overline{)20}$	$6\overline{)10}$	$2\overline{)9}$	$4\overline{)30}$	$8\overline{)15}$	$9\overline{)29}$
$5\overline{)32}$	$3\overline{)14}$	$9\overline{)50}$	$8\overline{)65}$	$2\overline{)11}$	$5\overline{)19}$

G | 48 Uneven Divisions
For use with Test 13

Name _____

Time _____

Divide. Write each answer with a remainder.

4)15	9)14	7)45	3)16	6)38	2)7
8)50	5)28	4)21	6)15	7)11	8)20
3)20	7)32	8)30	2)15	5)43	6)35
9)62	4)10	6)27	9)21	4)19	3)25
6)56	2)17	3)10	5)8	9)40	7)30
2)5	8)25	5)17	7)17	3)8	4)9
7)20	6)10	2)9	4)30	8)15	9)29
5)32	3)14	9)50	8)65	2)11	5)19

C 100 Multiplication Facts
For use with Lesson 71

Name _____

Time _____

Multiply.

9 × 9	3 × 5	8 × 5	2 × 6	4 × 7	0 × 3	7 × 2	1 × 5	7 × 8	4 × 0
3 × 4	5 × 9	0 × 2	7 × 3	4 × 1	2 × 7	6 × 3	5 × 4	1 × 0	9 × 2
1 × 1	9 × 0	2 × 8	6 × 4	0 × 7	8 × 1	3 × 3	4 × 8	9 × 3	2 × 0
4 × 9	7 × 0	1 × 2	8 × 4	6 × 5	2 × 9	9 × 4	0 × 1	7 × 4	5 × 8
0 × 8	4 × 2	9 × 8	3 × 6	5 × 5	1 × 6	5 × 0	6 × 6	2 × 1	7 × 9
9 × 1	2 × 2	5 × 1	4 × 3	0 × 0	8 × 9	3 × 7	9 × 7	1 × 7	6 × 0
5 × 6	7 × 5	3 × 0	8 × 8	1 × 3	8 × 3	5 × 2	0 × 4	9 × 5	6 × 7
2 × 3	8 × 6	0 × 5	6 × 1	3 × 8	7 × 6	1 × 8	9 × 6	4 × 4	5 × 3
7 × 7	1 × 4	6 × 2	4 × 5	2 × 4	8 × 0	3 × 1	6 × 8	0 × 9	8 × 7
3 × 2	4 × 6	1 × 9	5 × 7	8 × 2	0 × 6	7 × 1	2 × 5	6 × 9	3 × 9

E

90 Division Facts
For use with Lesson 72

Name _____

Time _____

Divide.

20 ÷ 4 =	21 ÷ 7 =	0 ÷ 2 =	27 ÷ 3 =	8 ÷ 1 =	54 ÷ 6 =
15 ÷ 5 =	6 ÷ 3 =	28 ÷ 4 =	18 ÷ 2 =	24 ÷ 6 =	9 ÷ 9 =
56 ÷ 8 =	0 ÷ 6 =	21 ÷ 3 =	1 ÷ 1 =	25 ÷ 5 =	12 ÷ 2 =
5 ÷ 1 =	45 ÷ 9 =	16 ÷ 4 =	30 ÷ 6 =	9 ÷ 3 =	14 ÷ 7 =
0 ÷ 8 =	6 ÷ 2 =	24 ÷ 8 =	10 ÷ 5 =	81 ÷ 9 =	24 ÷ 4 =
16 ÷ 2 =	30 ÷ 5 =	0 ÷ 1 =	28 ÷ 7 =	4 ÷ 4 =	40 ÷ 8 =
3 ÷ 3 =	18 ÷ 6 =	63 ÷ 9 =	40 ÷ 5 =	10 ÷ 2 =	36 ÷ 6 =
32 ÷ 8 =	12 ÷ 4 =	18 ÷ 3 =	35 ÷ 7 =	8 ÷ 8 =	2 ÷ 1 =
45 ÷ 5 =	7 ÷ 7 =	27 ÷ 9 =	9 ÷ 1 =	48 ÷ 6 =	0 ÷ 7 =
4 ÷ 1 =	0 ÷ 9 =	24 ÷ 3 =	32 ÷ 4 =	5 ÷ 5 =	72 ÷ 9 =
56 ÷ 7 =	15 ÷ 3 =	12 ÷ 6 =	8 ÷ 2 =	63 ÷ 7 =	0 ÷ 4 =
14 ÷ 2 =	42 ÷ 6 =	6 ÷ 1 =	16 ÷ 8 =	20 ÷ 5 =	49 ÷ 7 =
36 ÷ 4 =	64 ÷ 8 =	0 ÷ 3 =	54 ÷ 9 =	4 ÷ 2 =	48 ÷ 8 =
18 ÷ 9 =	3 ÷ 1 =	35 ÷ 5 =	8 ÷ 4 =	72 ÷ 8 =	6 ÷ 6 =
0 ÷ 5 =	42 ÷ 7 =	2 ÷ 2 =	36 ÷ 9 =	7 ÷ 1 =	12 ÷ 3 =

F

64 Multiplication Facts
For use with Lesson 73

Name _____

Time _____

Multiply.

5 × 6	4 × 3	9 × 8	7 × 5	2 × 9	8 × 4	9 × 3	6 × 9
9 × 4	2 × 5	7 × 6	4 × 8	7 × 9	5 × 4	3 × 2	9 × 7
3 × 7	8 × 5	6 × 2	5 × 5	3 × 5	2 × 4	7 × 7	8 × 9
6 × 4	2 × 8	4 × 4	8 × 2	3 × 9	6 × 6	9 × 9	5 × 3
4 × 6	8 × 8	5 × 7	6 × 3	2 × 2	7 × 4	3 × 8	8 × 6
2 × 6	5 × 9	3 × 3	9 × 2	6 × 7	4 × 5	7 × 2	9 × 6
5 × 2	7 × 8	2 × 3	6 × 8	4 × 7	9 × 5	3 × 6	8 × 7
3 × 4	7 × 3	5 × 8	4 × 2	8 × 3	2 × 7	6 × 5	4 × 9

G | **48 Uneven Divisions**
For use with Lesson 74

Name _____

Time _____

Divide. Write each answer with a remainder.

$4\overline{)15}$	$9\overline{)14}$	$7\overline{)45}$	$3\overline{)16}$	$6\overline{)38}$	$2\overline{)7}$
$8\overline{)50}$	$5\overline{)28}$	$4\overline{)21}$	$6\overline{)15}$	$7\overline{)11}$	$8\overline{)20}$
$3\overline{)20}$	$7\overline{)32}$	$8\overline{)30}$	$2\overline{)15}$	$5\overline{)43}$	$6\overline{)35}$
$9\overline{)62}$	$4\overline{)10}$	$6\overline{)27}$	$9\overline{)21}$	$4\overline{)19}$	$3\overline{)25}$
$6\overline{)56}$	$2\overline{)17}$	$3\overline{)10}$	$5\overline{)8}$	$9\overline{)40}$	$7\overline{)30}$
$2\overline{)5}$	$8\overline{)25}$	$5\overline{)17}$	$7\overline{)17}$	$3\overline{)8}$	$4\overline{)9}$
$7\overline{)20}$	$6\overline{)10}$	$2\overline{)9}$	$4\overline{)30}$	$8\overline{)15}$	$9\overline{)29}$
$5\overline{)32}$	$3\overline{)14}$	$9\overline{)50}$	$8\overline{)65}$	$2\overline{)11}$	$5\overline{)19}$

C

100 Multiplication Facts
For use with Lesson 75

Name _____

Time _____

Multiply.

9 × 9	3 × 5	8 × 5	2 × 6	4 × 7	0 × 3	7 × 2	1 × 5	7 × 8	4 × 0
3 × 4	5 × 9	0 × 2	7 × 3	4 × 1	2 × 7	6 × 3	5 × 4	1 × 0	9 × 2
1 × 1	9 × 0	2 × 8	6 × 4	0 × 7	8 × 1	3 × 3	4 × 8	9 × 3	2 × 0
4 × 9	7 × 0	1 × 2	8 × 4	6 × 5	2 × 9	9 × 4	0 × 1	7 × 4	5 × 8
0 × 8	4 × 2	9 × 8	3 × 6	5 × 5	1 × 6	5 × 0	6 × 6	2 × 1	7 × 9
9 × 1	2 × 2	5 × 1	4 × 3	0 × 0	8 × 9	3 × 7	9 × 7	1 × 7	6 × 0
5 × 6	7 × 5	3 × 0	8 × 8	1 × 3	8 × 3	5 × 2	0 × 4	9 × 5	6 × 7
2 × 3	8 × 6	0 × 5	6 × 1	3 × 8	7 × 6	1 × 8	9 × 6	4 × 4	5 × 3
7 × 7	1 × 4	6 × 2	4 × 5	2 × 4	8 × 0	3 × 1	6 × 8	0 × 9	8 × 7
3 × 2	4 × 6	1 × 9	5 × 7	8 × 2	0 × 6	7 × 1	2 × 5	6 × 9	3 × 9

C | **100 Multiplication Facts**
For use with Test 14

Name _____

Time _____

Multiply.

9 × 9	3 × 5	8 × 5	2 × 6	4 × 7	0 × 3	7 × 2	1 × 5	7 × 8	4 × 0
3 × 4	5 × 9	0 × 2	7 × 3	4 × 1	2 × 7	6 × 3	5 × 4	1 × 0	9 × 2
1 × 1	9 × 0	2 × 8	6 × 4	0 × 7	8 × 1	3 × 3	4 × 8	9 × 3	2 × 0
4 × 9	7 × 0	1 × 2	8 × 4	6 × 5	2 × 9	9 × 4	0 × 1	7 × 4	5 × 8
0 × 8	4 × 2	9 × 8	3 × 6	5 × 5	1 × 6	5 × 0	6 × 6	2 × 1	7 × 9
9 × 1	2 × 2	5 × 1	4 × 3	0 × 0	8 × 9	3 × 7	9 × 7	1 × 7	6 × 0
5 × 6	7 × 5	3 × 0	8 × 8	1 × 3	8 × 3	5 × 2	0 × 4	9 × 5	6 × 7
2 × 3	8 × 6	0 × 5	6 × 1	3 × 8	7 × 6	1 × 8	9 × 6	4 × 4	5 × 3
7 × 7	1 × 4	6 × 2	4 × 5	2 × 4	8 × 0	3 × 1	6 × 8	0 × 9	8 × 7
3 × 2	4 × 6	1 × 9	5 × 7	8 × 2	0 × 6	7 × 1	2 × 5	6 × 9	3 × 9

FACTS PRACTICE TEST

H	**60 Improper Fractions to Simplify**

For use with Lesson 76

Name _____

Time _____

Simplify.

$\frac{15}{2} =$	$\frac{9}{8} =$	$\frac{10}{2} =$	$\frac{18}{6} =$	$\frac{8}{3} =$	$\frac{12}{4} =$
$\frac{10}{10} =$	$\frac{3}{2} =$	$\frac{11}{4} =$	$\frac{4}{3} =$	$\frac{12}{5} =$	$\frac{5}{4} =$
$\frac{12}{6} =$	$\frac{9}{3} =$	$\frac{5}{5} =$	$\frac{15}{4} =$	$\frac{6}{2} =$	$\frac{9}{9} =$
$\frac{3}{3} =$	$\frac{7}{4} =$	$\frac{21}{10} =$	$\frac{11}{2} =$	$\frac{7}{6} =$	$\frac{24}{8} =$
$\frac{11}{3} =$	$\frac{9}{5} =$	$\frac{4}{2} =$	$\frac{21}{8} =$	$\frac{6}{5} =$	$\frac{12}{3} =$
$\frac{7}{2} =$	$\frac{25}{6} =$	$\frac{10}{9} =$	$\frac{4}{4} =$	$\frac{12}{2} =$	$\frac{16}{15} =$
$\frac{10}{5} =$	$\frac{5}{2} =$	$\frac{7}{3} =$	$\frac{8}{4} =$	$\frac{8}{8} =$	$\frac{27}{10} =$
$\frac{16}{4} =$	$\frac{6}{6} =$	$\frac{25}{12} =$	$\frac{5}{3} =$	$\frac{7}{5} =$	$\frac{16}{9} =$
$\frac{15}{8} =$	$\frac{10}{3} =$	$\frac{33}{10} =$	$\frac{2}{2} =$	$\frac{35}{6} =$	$\frac{25}{8} =$
$\frac{6}{3} =$	$\frac{8}{5} =$	$\frac{9}{4} =$	$\frac{12}{12} =$	$\frac{25}{2} =$	$\frac{9}{2} =$

H

60 Improper Fractions to Simplify
For use with Lesson 77

Name _____

Time _____

Simplify.

$\frac{15}{2} =$	$\frac{9}{8} =$	$\frac{10}{2} =$	$\frac{18}{6} =$	$\frac{8}{3} =$	$\frac{12}{4} =$
$\frac{10}{10} =$	$\frac{3}{2} =$	$\frac{11}{4} =$	$\frac{4}{3} =$	$\frac{12}{5} =$	$\frac{5}{4} =$
$\frac{12}{6} =$	$\frac{9}{3} =$	$\frac{5}{5} =$	$\frac{15}{4} =$	$\frac{6}{2} =$	$\frac{9}{9} =$
$\frac{3}{3} =$	$\frac{7}{4} =$	$\frac{21}{10} =$	$\frac{11}{2} =$	$\frac{7}{6} =$	$\frac{24}{8} =$
$\frac{11}{3} =$	$\frac{9}{5} =$	$\frac{4}{2} =$	$\frac{21}{8} =$	$\frac{6}{5} =$	$\frac{12}{3} =$
$\frac{7}{2} =$	$\frac{25}{6} =$	$\frac{10}{9} =$	$\frac{4}{4} =$	$\frac{12}{2} =$	$\frac{16}{15} =$
$\frac{10}{5} =$	$\frac{5}{2} =$	$\frac{7}{3} =$	$\frac{8}{4} =$	$\frac{8}{8} =$	$\frac{27}{10} =$
$\frac{16}{4} =$	$\frac{6}{6} =$	$\frac{25}{12} =$	$\frac{5}{3} =$	$\frac{7}{5} =$	$\frac{16}{9} =$
$\frac{15}{8} =$	$\frac{10}{3} =$	$\frac{33}{10} =$	$\frac{2}{2} =$	$\frac{35}{6} =$	$\frac{25}{8} =$
$\frac{6}{3} =$	$\frac{8}{5} =$	$\frac{9}{4} =$	$\frac{12}{12} =$	$\frac{25}{2} =$	$\frac{9}{2} =$

60 Improper Fractions to Simplify
For use with Lesson 78

Name _____

Time _____

Simplify.

$\frac{15}{2} =$	$\frac{9}{8} =$	$\frac{10}{2} =$	$\frac{18}{6} =$	$\frac{8}{3} =$	$\frac{12}{4} =$
$\frac{10}{10} =$	$\frac{3}{2} =$	$\frac{11}{4} =$	$\frac{4}{3} =$	$\frac{12}{5} =$	$\frac{5}{4} =$
$\frac{12}{6} =$	$\frac{9}{3} =$	$\frac{5}{5} =$	$\frac{15}{4} =$	$\frac{6}{2} =$	$\frac{9}{9} =$
$\frac{3}{3} =$	$\frac{7}{4} =$	$\frac{21}{10} =$	$\frac{11}{2} =$	$\frac{7}{6} =$	$\frac{24}{8} =$
$\frac{11}{3} =$	$\frac{9}{5} =$	$\frac{4}{2} =$	$\frac{21}{8} =$	$\frac{6}{5} =$	$\frac{12}{3} =$
$\frac{7}{2} =$	$\frac{25}{6} =$	$\frac{10}{9} =$	$\frac{4}{4} =$	$\frac{12}{2} =$	$\frac{16}{15} =$
$\frac{10}{5} =$	$\frac{5}{2} =$	$\frac{7}{3} =$	$\frac{8}{4} =$	$\frac{8}{8} =$	$\frac{27}{10} =$
$\frac{16}{4} =$	$\frac{6}{6} =$	$\frac{25}{12} =$	$\frac{5}{3} =$	$\frac{7}{5} =$	$\frac{16}{9} =$
$\frac{15}{8} =$	$\frac{10}{3} =$	$\frac{33}{10} =$	$\frac{2}{2} =$	$\frac{35}{6} =$	$\frac{25}{8} =$
$\frac{6}{3} =$	$\frac{8}{5} =$	$\frac{9}{4} =$	$\frac{12}{12} =$	$\frac{25}{2} =$	$\frac{9}{2} =$

H

60 Improper Fractions to Simplify

For use with Lesson 79

Name _____

Time _____

Simplify.

$\frac{15}{2} =$	$\frac{9}{8} =$	$\frac{10}{2} =$	$\frac{18}{6} =$	$\frac{8}{3} =$	$\frac{12}{4} =$
$\frac{10}{10} =$	$\frac{3}{2} =$	$\frac{11}{4} =$	$\frac{4}{3} =$	$\frac{12}{5} =$	$\frac{5}{4} =$
$\frac{12}{6} =$	$\frac{9}{3} =$	$\frac{5}{5} =$	$\frac{15}{4} =$	$\frac{6}{2} =$	$\frac{9}{9} =$
$\frac{3}{3} =$	$\frac{7}{4} =$	$\frac{21}{10} =$	$\frac{11}{2} =$	$\frac{7}{6} =$	$\frac{24}{8} =$
$\frac{11}{3} =$	$\frac{9}{5} =$	$\frac{4}{2} =$	$\frac{21}{8} =$	$\frac{6}{5} =$	$\frac{12}{3} =$
$\frac{7}{2} =$	$\frac{25}{6} =$	$\frac{10}{9} =$	$\frac{4}{4} =$	$\frac{12}{2} =$	$\frac{16}{15} =$
$\frac{10}{5} =$	$\frac{5}{2} =$	$\frac{7}{3} =$	$\frac{8}{4} =$	$\frac{8}{8} =$	$\frac{27}{10} =$
$\frac{16}{4} =$	$\frac{6}{6} =$	$\frac{25}{12} =$	$\frac{5}{3} =$	$\frac{7}{5} =$	$\frac{16}{9} =$
$\frac{15}{8} =$	$\frac{10}{3} =$	$\frac{33}{10} =$	$\frac{2}{2} =$	$\frac{35}{6} =$	$\frac{25}{8} =$
$\frac{6}{3} =$	$\frac{8}{5} =$	$\frac{9}{4} =$	$\frac{12}{12} =$	$\frac{25}{2} =$	$\frac{9}{2} =$

60 Improper Fractions to Simplify
For use with Lesson 80

Name _____

Time _____

Simplify.

$\frac{15}{2} =$	$\frac{9}{8} =$	$\frac{10}{2} =$	$\frac{18}{6} =$	$\frac{8}{3} =$	$\frac{12}{4} =$
$\frac{10}{10} =$	$\frac{3}{2} =$	$\frac{11}{4} =$	$\frac{4}{3} =$	$\frac{12}{5} =$	$\frac{5}{4} =$
$\frac{12}{6} =$	$\frac{9}{3} =$	$\frac{5}{5} =$	$\frac{15}{4} =$	$\frac{6}{2} =$	$\frac{9}{9} =$
$\frac{3}{3} =$	$\frac{7}{4} =$	$\frac{21}{10} =$	$\frac{11}{2} =$	$\frac{7}{6} =$	$\frac{24}{8} =$
$\frac{11}{3} =$	$\frac{9}{5} =$	$\frac{4}{2} =$	$\frac{21}{8} =$	$\frac{6}{5} =$	$\frac{12}{3} =$
$\frac{7}{2} =$	$\frac{25}{6} =$	$\frac{10}{9} =$	$\frac{4}{4} =$	$\frac{12}{2} =$	$\frac{16}{15} =$
$\frac{10}{5} =$	$\frac{5}{2} =$	$\frac{7}{3} =$	$\frac{8}{4} =$	$\frac{8}{8} =$	$\frac{27}{10} =$
$\frac{16}{4} =$	$\frac{6}{6} =$	$\frac{25}{12} =$	$\frac{5}{3} =$	$\frac{7}{5} =$	$\frac{16}{9} =$
$\frac{15}{8} =$	$\frac{10}{3} =$	$\frac{33}{10} =$	$\frac{2}{2} =$	$\frac{35}{6} =$	$\frac{25}{8} =$
$\frac{6}{3} =$	$\frac{8}{5} =$	$\frac{9}{4} =$	$\frac{12}{12} =$	$\frac{25}{2} =$	$\frac{9}{2} =$

H

60 Improper Fractions to Simplify
For use with Test 15

Name _____

Time _____

Simplify.

$\frac{15}{2} =$	$\frac{9}{8} =$	$\frac{10}{2} =$	$\frac{18}{6} =$	$\frac{8}{3} =$	$\frac{12}{4} =$
$\frac{10}{10} =$	$\frac{3}{2} =$	$\frac{11}{4} =$	$\frac{4}{3} =$	$\frac{12}{5} =$	$\frac{5}{4} =$
$\frac{12}{6} =$	$\frac{9}{3} =$	$\frac{5}{5} =$	$\frac{15}{4} =$	$\frac{6}{2} =$	$\frac{9}{9} =$
$\frac{3}{3} =$	$\frac{7}{4} =$	$\frac{21}{10} =$	$\frac{11}{2} =$	$\frac{7}{6} =$	$\frac{24}{8} =$
$\frac{11}{3} =$	$\frac{9}{5} =$	$\frac{4}{2} =$	$\frac{21}{8} =$	$\frac{6}{5} =$	$\frac{12}{3} =$
$\frac{7}{2} =$	$\frac{25}{6} =$	$\frac{10}{9} =$	$\frac{4}{4} =$	$\frac{12}{2} =$	$\frac{16}{15} =$
$\frac{10}{5} =$	$\frac{5}{2} =$	$\frac{7}{3} =$	$\frac{8}{4} =$	$\frac{8}{8} =$	$\frac{27}{10} =$
$\frac{16}{4} =$	$\frac{6}{6} =$	$\frac{25}{12} =$	$\frac{5}{3} =$	$\frac{7}{5} =$	$\frac{16}{9} =$
$\frac{15}{8} =$	$\frac{10}{3} =$	$\frac{33}{10} =$	$\frac{2}{2} =$	$\frac{35}{6} =$	$\frac{25}{8} =$
$\frac{6}{3} =$	$\frac{8}{5} =$	$\frac{9}{4} =$	$\frac{12}{12} =$	$\frac{25}{2} =$	$\frac{9}{2} =$

60 Improper Fractions to Simplify
For use with Lesson 81

Name _____

Time _____

Simplify.

$\frac{15}{2}$ =	$\frac{9}{8}$ =	$\frac{10}{2}$ =	$\frac{18}{6}$ =	$\frac{8}{3}$ =	$\frac{12}{4}$ =
$\frac{10}{10}$ =	$\frac{3}{2}$ =	$\frac{11}{4}$ =	$\frac{4}{3}$ =	$\frac{12}{5}$ =	$\frac{5}{4}$ =
$\frac{12}{6}$ =	$\frac{9}{3}$ =	$\frac{5}{5}$ =	$\frac{15}{4}$ =	$\frac{6}{2}$ =	$\frac{9}{9}$ =
$\frac{3}{3}$ =	$\frac{7}{4}$ =	$\frac{21}{10}$ =	$\frac{11}{2}$ =	$\frac{7}{6}$ =	$\frac{24}{8}$ =
$\frac{11}{3}$ =	$\frac{9}{5}$ =	$\frac{4}{2}$ =	$\frac{21}{8}$ =	$\frac{6}{5}$ =	$\frac{12}{3}$ =
$\frac{7}{2}$ =	$\frac{25}{6}$ =	$\frac{10}{9}$ =	$\frac{4}{4}$ =	$\frac{12}{2}$ =	$\frac{16}{15}$ =
$\frac{10}{5}$ =	$\frac{5}{2}$ =	$\frac{7}{3}$ =	$\frac{8}{4}$ =	$\frac{8}{8}$ =	$\frac{27}{10}$ =
$\frac{16}{4}$ =	$\frac{6}{6}$ =	$\frac{25}{12}$ =	$\frac{5}{3}$ =	$\frac{7}{5}$ =	$\frac{16}{9}$ =
$\frac{15}{8}$ =	$\frac{10}{3}$ =	$\frac{33}{10}$ =	$\frac{2}{2}$ =	$\frac{35}{6}$ =	$\frac{25}{8}$ =
$\frac{6}{3}$ =	$\frac{8}{5}$ =	$\frac{9}{4}$ =	$\frac{12}{12}$ =	$\frac{25}{2}$ =	$\frac{9}{2}$ =

H
60 Improper Fractions to Simplify
For use with Lesson 82

Name _____

Time _____

Simplify.

$\frac{15}{2} =$	$\frac{9}{8} =$	$\frac{10}{2} =$	$\frac{18}{6} =$	$\frac{8}{3} =$	$\frac{12}{4} =$
$\frac{10}{10} =$	$\frac{3}{2} =$	$\frac{11}{4} =$	$\frac{4}{3} =$	$\frac{12}{5} =$	$\frac{5}{4} =$
$\frac{12}{6} =$	$\frac{9}{3} =$	$\frac{5}{5} =$	$\frac{15}{4} =$	$\frac{6}{2} =$	$\frac{9}{9} =$
$\frac{3}{3} =$	$\frac{7}{4} =$	$\frac{21}{10} =$	$\frac{11}{2} =$	$\frac{7}{6} =$	$\frac{24}{8} =$
$\frac{11}{3} =$	$\frac{9}{5} =$	$\frac{4}{2} =$	$\frac{21}{8} =$	$\frac{6}{5} =$	$\frac{12}{3} =$
$\frac{7}{2} =$	$\frac{25}{6} =$	$\frac{10}{9} =$	$\frac{4}{4} =$	$\frac{12}{2} =$	$\frac{16}{15} =$
$\frac{10}{5} =$	$\frac{5}{2} =$	$\frac{7}{3} =$	$\frac{8}{4} =$	$\frac{8}{8} =$	$\frac{27}{10} =$
$\frac{16}{4} =$	$\frac{6}{6} =$	$\frac{25}{12} =$	$\frac{5}{3} =$	$\frac{7}{5} =$	$\frac{16}{9} =$
$\frac{15}{8} =$	$\frac{10}{3} =$	$\frac{33}{10} =$	$\frac{2}{2} =$	$\frac{35}{6} =$	$\frac{25}{8} =$
$\frac{6}{3} =$	$\frac{8}{5} =$	$\frac{9}{4} =$	$\frac{12}{12} =$	$\frac{25}{2} =$	$\frac{9}{2} =$

H	**60 Improper Fractions to Simplify**

For use with Lesson 83

Name _____

Time _____

Simplify.

$\dfrac{15}{2} =$	$\dfrac{9}{8} =$	$\dfrac{10}{2} =$	$\dfrac{18}{6} =$	$\dfrac{8}{3} =$	$\dfrac{12}{4} =$
$\dfrac{10}{10} =$	$\dfrac{3}{2} =$	$\dfrac{11}{4} =$	$\dfrac{4}{3} =$	$\dfrac{12}{5} =$	$\dfrac{5}{4} =$
$\dfrac{12}{6} =$	$\dfrac{9}{3} =$	$\dfrac{5}{5} =$	$\dfrac{15}{4} =$	$\dfrac{6}{2} =$	$\dfrac{9}{9} =$
$\dfrac{3}{3} =$	$\dfrac{7}{4} =$	$\dfrac{21}{10} =$	$\dfrac{11}{2} =$	$\dfrac{7}{6} =$	$\dfrac{24}{8} =$
$\dfrac{11}{3} =$	$\dfrac{9}{5} =$	$\dfrac{4}{2} =$	$\dfrac{21}{8} =$	$\dfrac{6}{5} =$	$\dfrac{12}{3} =$
$\dfrac{7}{2} =$	$\dfrac{25}{6} =$	$\dfrac{10}{9} =$	$\dfrac{4}{4} =$	$\dfrac{12}{2} =$	$\dfrac{16}{15} =$
$\dfrac{10}{5} =$	$\dfrac{5}{2} =$	$\dfrac{7}{3} =$	$\dfrac{8}{4} =$	$\dfrac{8}{8} =$	$\dfrac{27}{10} =$
$\dfrac{16}{4} =$	$\dfrac{6}{6} =$	$\dfrac{25}{12} =$	$\dfrac{5}{3} =$	$\dfrac{7}{5} =$	$\dfrac{16}{9} =$
$\dfrac{15}{8} =$	$\dfrac{10}{3} =$	$\dfrac{33}{10} =$	$\dfrac{2}{2} =$	$\dfrac{35}{6} =$	$\dfrac{25}{8} =$
$\dfrac{6}{3} =$	$\dfrac{8}{5} =$	$\dfrac{9}{4} =$	$\dfrac{12}{12} =$	$\dfrac{25}{2} =$	$\dfrac{9}{2} =$

H	**60 Improper Fractions to Simplify**

For use with Lesson 84

Name _____

Time _____

Simplify.

$\frac{15}{2} =$	$\frac{9}{8} =$	$\frac{10}{2} =$	$\frac{18}{6} =$	$\frac{8}{3} =$	$\frac{12}{4} =$
$\frac{10}{10} =$	$\frac{3}{2} =$	$\frac{11}{4} =$	$\frac{4}{3} =$	$\frac{12}{5} =$	$\frac{5}{4} =$
$\frac{12}{6} =$	$\frac{9}{3} =$	$\frac{5}{5} =$	$\frac{15}{4} =$	$\frac{6}{2} =$	$\frac{9}{9} =$
$\frac{3}{3} =$	$\frac{7}{4} =$	$\frac{21}{10} =$	$\frac{11}{2} =$	$\frac{7}{6} =$	$\frac{24}{8} =$
$\frac{11}{3} =$	$\frac{9}{5} =$	$\frac{4}{2} =$	$\frac{21}{8} =$	$\frac{6}{5} =$	$\frac{12}{3} =$
$\frac{7}{2} =$	$\frac{25}{6} =$	$\frac{10}{9} =$	$\frac{4}{4} =$	$\frac{12}{2} =$	$\frac{16}{15} =$
$\frac{10}{5} =$	$\frac{5}{2} =$	$\frac{7}{3} =$	$\frac{8}{4} =$	$\frac{8}{8} =$	$\frac{27}{10} =$
$\frac{16}{4} =$	$\frac{6}{6} =$	$\frac{25}{12} =$	$\frac{5}{3} =$	$\frac{7}{5} =$	$\frac{16}{9} =$
$\frac{15}{8} =$	$\frac{10}{3} =$	$\frac{33}{10} =$	$\frac{2}{2} =$	$\frac{35}{6} =$	$\frac{25}{8} =$
$\frac{6}{3} =$	$\frac{8}{5} =$	$\frac{9}{4} =$	$\frac{12}{12} =$	$\frac{25}{2} =$	$\frac{9}{2} =$

**60 Improper Fractions
to Simplify**
For use with Lesson 85

Name _____

Time _____

Simplify.

$\frac{15}{2}$ =	$\frac{9}{8}$ =	$\frac{10}{2}$ =	$\frac{18}{6}$ =	$\frac{8}{3}$ =	$\frac{12}{4}$ =
$\frac{10}{10}$ =	$\frac{3}{2}$ =	$\frac{11}{4}$ =	$\frac{4}{3}$ =	$\frac{12}{5}$ =	$\frac{5}{4}$ =
$\frac{12}{6}$ =	$\frac{9}{3}$ =	$\frac{5}{5}$ =	$\frac{15}{4}$ =	$\frac{6}{2}$ =	$\frac{9}{9}$ =
$\frac{3}{3}$ =	$\frac{7}{4}$ =	$\frac{21}{10}$ =	$\frac{11}{2}$ =	$\frac{7}{6}$ =	$\frac{24}{8}$ =
$\frac{11}{3}$ =	$\frac{9}{5}$ =	$\frac{4}{2}$ =	$\frac{21}{8}$ =	$\frac{6}{5}$ =	$\frac{12}{3}$ =
$\frac{7}{2}$ =	$\frac{25}{6}$ =	$\frac{10}{9}$ =	$\frac{4}{4}$ =	$\frac{12}{2}$ =	$\frac{16}{15}$ =
$\frac{10}{5}$ =	$\frac{5}{2}$ =	$\frac{7}{3}$ =	$\frac{8}{4}$ =	$\frac{8}{8}$ =	$\frac{27}{10}$ =
$\frac{16}{4}$ =	$\frac{6}{6}$ =	$\frac{25}{12}$ =	$\frac{5}{3}$ =	$\frac{7}{5}$ =	$\frac{16}{9}$ =
$\frac{15}{8}$ =	$\frac{10}{3}$ =	$\frac{33}{10}$ =	$\frac{2}{2}$ =	$\frac{35}{6}$ =	$\frac{25}{8}$ =
$\frac{6}{3}$ =	$\frac{8}{5}$ =	$\frac{9}{4}$ =	$\frac{12}{12}$ =	$\frac{25}{2}$ =	$\frac{9}{2}$ =

H

60 Improper Fractions to Simplify
For use with Test 16

Name _____

Time _____

Simplify.

$\frac{15}{2} =$	$\frac{9}{8} =$	$\frac{10}{2} =$	$\frac{18}{6} =$	$\frac{8}{3} =$	$\frac{12}{4} =$
$\frac{10}{10} =$	$\frac{3}{2} =$	$\frac{11}{4} =$	$\frac{4}{3} =$	$\frac{12}{5} =$	$\frac{5}{4} =$
$\frac{12}{6} =$	$\frac{9}{3} =$	$\frac{5}{5} =$	$\frac{15}{4} =$	$\frac{6}{2} =$	$\frac{9}{9} =$
$\frac{3}{3} =$	$\frac{7}{4} =$	$\frac{21}{10} =$	$\frac{11}{2} =$	$\frac{7}{6} =$	$\frac{24}{8} =$
$\frac{11}{3} =$	$\frac{9}{5} =$	$\frac{4}{2} =$	$\frac{21}{8} =$	$\frac{6}{5} =$	$\frac{12}{3} =$
$\frac{7}{2} =$	$\frac{25}{6} =$	$\frac{10}{9} =$	$\frac{4}{4} =$	$\frac{12}{2} =$	$\frac{16}{15} =$
$\frac{10}{5} =$	$\frac{5}{2} =$	$\frac{7}{3} =$	$\frac{8}{4} =$	$\frac{8}{8} =$	$\frac{27}{10} =$
$\frac{16}{4} =$	$\frac{6}{6} =$	$\frac{25}{12} =$	$\frac{5}{3} =$	$\frac{7}{5} =$	$\frac{16}{9} =$
$\frac{15}{8} =$	$\frac{10}{3} =$	$\frac{33}{10} =$	$\frac{2}{2} =$	$\frac{35}{6} =$	$\frac{25}{8} =$
$\frac{6}{3} =$	$\frac{8}{5} =$	$\frac{9}{4} =$	$\frac{12}{12} =$	$\frac{25}{2} =$	$\frac{9}{2} =$

FACTS PRACTICE TEST

H

60 Improper Fractions to Simplify
For use with Lesson 86

Name _____

Time _____

Simplify.

$\frac{15}{2} =$	$\frac{9}{8} =$	$\frac{10}{2} =$	$\frac{18}{6} =$	$\frac{8}{3} =$	$\frac{12}{4} =$
$\frac{10}{10} =$	$\frac{3}{2} =$	$\frac{11}{4} =$	$\frac{4}{3} =$	$\frac{12}{5} =$	$\frac{5}{4} =$
$\frac{12}{6} =$	$\frac{9}{3} =$	$\frac{5}{5} =$	$\frac{15}{4} =$	$\frac{6}{2} =$	$\frac{9}{9} =$
$\frac{3}{3} =$	$\frac{7}{4} =$	$\frac{21}{10} =$	$\frac{11}{2} =$	$\frac{7}{6} =$	$\frac{24}{8} =$
$\frac{11}{3} =$	$\frac{9}{5} =$	$\frac{4}{2} =$	$\frac{21}{8} =$	$\frac{6}{5} =$	$\frac{12}{3} =$
$\frac{7}{2} =$	$\frac{25}{6} =$	$\frac{10}{9} =$	$\frac{4}{4} =$	$\frac{12}{2} =$	$\frac{16}{15} =$
$\frac{10}{5} =$	$\frac{5}{2} =$	$\frac{7}{3} =$	$\frac{8}{4} =$	$\frac{8}{8} =$	$\frac{27}{10} =$
$\frac{16}{4} =$	$\frac{6}{6} =$	$\frac{25}{12} =$	$\frac{5}{3} =$	$\frac{7}{5} =$	$\frac{16}{9} =$
$\frac{15}{8} =$	$\frac{10}{3} =$	$\frac{33}{10} =$	$\frac{2}{2} =$	$\frac{35}{6} =$	$\frac{25}{8} =$
$\frac{6}{3} =$	$\frac{8}{5} =$	$\frac{9}{4} =$	$\frac{12}{12} =$	$\frac{25}{2} =$	$\frac{9}{2} =$

H	**60 Improper Fractions to Simplify**

For use with Lesson 87

Name _____

Time _____

Simplify.

$\dfrac{15}{2}=$	$\dfrac{9}{8}=$	$\dfrac{10}{2}=$	$\dfrac{18}{6}=$	$\dfrac{8}{3}=$	$\dfrac{12}{4}=$
$\dfrac{10}{10}=$	$\dfrac{3}{2}=$	$\dfrac{11}{4}=$	$\dfrac{4}{3}=$	$\dfrac{12}{5}=$	$\dfrac{5}{4}=$
$\dfrac{12}{6}=$	$\dfrac{9}{3}=$	$\dfrac{5}{5}=$	$\dfrac{15}{4}=$	$\dfrac{6}{2}=$	$\dfrac{9}{9}=$
$\dfrac{3}{3}=$	$\dfrac{7}{4}=$	$\dfrac{21}{10}=$	$\dfrac{11}{2}=$	$\dfrac{7}{6}=$	$\dfrac{24}{8}=$
$\dfrac{11}{3}=$	$\dfrac{9}{5}=$	$\dfrac{4}{2}=$	$\dfrac{21}{8}=$	$\dfrac{6}{5}=$	$\dfrac{12}{3}=$
$\dfrac{7}{2}=$	$\dfrac{25}{6}=$	$\dfrac{10}{9}=$	$\dfrac{4}{4}=$	$\dfrac{12}{2}=$	$\dfrac{16}{15}=$
$\dfrac{10}{5}=$	$\dfrac{5}{2}=$	$\dfrac{7}{3}=$	$\dfrac{8}{4}=$	$\dfrac{8}{8}=$	$\dfrac{27}{10}=$
$\dfrac{16}{4}=$	$\dfrac{6}{6}=$	$\dfrac{25}{12}=$	$\dfrac{5}{3}=$	$\dfrac{7}{5}=$	$\dfrac{16}{9}=$
$\dfrac{15}{8}=$	$\dfrac{10}{3}=$	$\dfrac{33}{10}=$	$\dfrac{2}{2}=$	$\dfrac{35}{6}=$	$\dfrac{25}{8}=$
$\dfrac{6}{3}=$	$\dfrac{8}{5}=$	$\dfrac{9}{4}=$	$\dfrac{12}{12}=$	$\dfrac{25}{2}=$	$\dfrac{9}{2}=$

H

60 Improper Fractions to Simplify
For use with Lesson 88

Name _____

Time _____

Simplify.

$\frac{15}{2}$ =	$\frac{9}{8}$ =	$\frac{10}{2}$ =	$\frac{18}{6}$ =	$\frac{8}{3}$ =	$\frac{12}{4}$ =
$\frac{10}{10}$ =	$\frac{3}{2}$ =	$\frac{11}{4}$ =	$\frac{4}{3}$ =	$\frac{12}{5}$ =	$\frac{5}{4}$ =
$\frac{12}{6}$ =	$\frac{9}{3}$ =	$\frac{5}{5}$ =	$\frac{15}{4}$ =	$\frac{6}{2}$ =	$\frac{9}{9}$ =
$\frac{3}{3}$ =	$\frac{7}{4}$ =	$\frac{21}{10}$ =	$\frac{11}{2}$ =	$\frac{7}{6}$ =	$\frac{24}{8}$ =
$\frac{11}{3}$ =	$\frac{9}{5}$ =	$\frac{4}{2}$ =	$\frac{21}{8}$ =	$\frac{6}{5}$ =	$\frac{12}{3}$ =
$\frac{7}{2}$ =	$\frac{25}{6}$ =	$\frac{10}{9}$ =	$\frac{4}{4}$ =	$\frac{12}{2}$ =	$\frac{16}{15}$ =
$\frac{10}{5}$ =	$\frac{5}{2}$ =	$\frac{7}{3}$ =	$\frac{8}{4}$ =	$\frac{8}{8}$ =	$\frac{27}{10}$ =
$\frac{16}{4}$ =	$\frac{6}{6}$ =	$\frac{25}{12}$ =	$\frac{5}{3}$ =	$\frac{7}{5}$ =	$\frac{16}{9}$ =
$\frac{15}{8}$ =	$\frac{10}{3}$ =	$\frac{33}{10}$ =	$\frac{2}{2}$ =	$\frac{35}{6}$ =	$\frac{25}{8}$ =
$\frac{6}{3}$ =	$\frac{8}{5}$ =	$\frac{9}{4}$ =	$\frac{12}{12}$ =	$\frac{25}{2}$ =	$\frac{9}{2}$ =

F | 64 Multiplication Facts

For use with Lesson 89

Name _____

Time _____

Multiply.

5 × 6	4 × 3	9 × 8	7 × 5	2 × 9	8 × 4	9 × 3	6 × 9
9 × 4	2 × 5	7 × 6	4 × 8	7 × 9	5 × 4	3 × 2	9 × 7
3 × 7	8 × 5	6 × 2	5 × 5	3 × 5	2 × 4	7 × 7	8 × 9
6 × 4	2 × 8	4 × 4	8 × 2	3 × 9	6 × 6	9 × 9	5 × 3
4 × 6	8 × 8	5 × 7	6 × 3	2 × 2	7 × 4	3 × 8	8 × 6
2 × 6	5 × 9	3 × 3	9 × 2	6 × 7	4 × 5	7 × 2	9 × 6
5 × 2	7 × 8	2 × 3	6 × 8	4 × 7	9 × 5	3 × 6	8 × 7
3 × 4	7 × 3	5 × 8	4 × 2	8 × 3	2 × 7	6 × 5	4 × 9

Saxon Math 6/5—Homeschool

F | 64 Multiplication Facts
For use with Lesson 90

Name _____

Time _____

Multiply.

5 × 6	4 × 3	9 × 8	7 × 5	2 × 9	8 × 4	9 × 3	6 × 9
9 × 4	2 × 5	7 × 6	4 × 8	7 × 9	5 × 4	3 × 2	9 × 7
3 × 7	8 × 5	6 × 2	5 × 5	3 × 5	2 × 4	7 × 7	8 × 9
6 × 4	2 × 8	4 × 4	8 × 2	3 × 9	6 × 6	9 × 9	5 × 3
4 × 6	8 × 8	5 × 7	6 × 3	2 × 2	7 × 4	3 × 8	8 × 6
2 × 6	5 × 9	3 × 3	9 × 2	6 × 7	4 × 5	7 × 2	9 × 6
5 × 2	7 × 8	2 × 3	6 × 8	4 × 7	9 × 5	3 × 6	8 × 7
3 × 4	7 × 3	5 × 8	4 × 2	8 × 3	2 × 7	6 × 5	4 × 9

60 Improper Fractions to Simplify
For use with Test 17

Name _____

Time _____

Simplify.

$\frac{15}{2}$ =	$\frac{9}{8}$ =	$\frac{10}{2}$ =	$\frac{18}{6}$ =	$\frac{8}{3}$ =	$\frac{12}{4}$ =
$\frac{10}{10}$ =	$\frac{3}{2}$ =	$\frac{11}{4}$ =	$\frac{4}{3}$ =	$\frac{12}{5}$ =	$\frac{5}{4}$ =
$\frac{12}{6}$ =	$\frac{9}{3}$ =	$\frac{5}{5}$ =	$\frac{15}{4}$ =	$\frac{6}{2}$ =	$\frac{9}{9}$ =
$\frac{3}{3}$ =	$\frac{7}{4}$ =	$\frac{21}{10}$ =	$\frac{11}{2}$ =	$\frac{7}{6}$ =	$\frac{24}{8}$ =
$\frac{11}{3}$ =	$\frac{9}{5}$ =	$\frac{4}{2}$ =	$\frac{21}{8}$ =	$\frac{6}{5}$ =	$\frac{12}{3}$ =
$\frac{7}{2}$ =	$\frac{25}{6}$ =	$\frac{10}{9}$ =	$\frac{4}{4}$ =	$\frac{12}{2}$ =	$\frac{16}{15}$ =
$\frac{10}{5}$ =	$\frac{5}{2}$ =	$\frac{7}{3}$ =	$\frac{8}{4}$ =	$\frac{8}{8}$ =	$\frac{27}{10}$ =
$\frac{16}{4}$ =	$\frac{6}{6}$ =	$\frac{25}{12}$ =	$\frac{5}{3}$ =	$\frac{7}{5}$ =	$\frac{16}{9}$ =
$\frac{15}{8}$ =	$\frac{10}{3}$ =	$\frac{33}{10}$ =	$\frac{2}{2}$ =	$\frac{35}{6}$ =	$\frac{25}{8}$ =
$\frac{6}{3}$ =	$\frac{8}{5}$ =	$\frac{9}{4}$ =	$\frac{12}{12}$ =	$\frac{25}{2}$ =	$\frac{9}{2}$ =

I

40 Fractions to Reduce
For use with Lesson 91

Name _____

Time _____

Reduce each fraction to lowest terms.

$\dfrac{2}{10} =$	$\dfrac{8}{16} =$	$\dfrac{2}{6} =$	$\dfrac{10}{100} =$	$\dfrac{6}{8} =$
$\dfrac{10}{15} =$	$\dfrac{5}{10} =$	$\dfrac{8}{12} =$	$\dfrac{9}{15} =$	$\dfrac{4}{16} =$
$\dfrac{2}{8} =$	$\dfrac{4}{10} =$	$\dfrac{15}{20} =$	$\dfrac{4}{8} =$	$\dfrac{4}{6} =$
$\dfrac{6}{15} =$	$\dfrac{4}{12} =$	$\dfrac{25}{100} =$	$\dfrac{10}{25} =$	$\dfrac{12}{20} =$
$\dfrac{20}{100} =$	$\dfrac{6}{9} =$	$\dfrac{2}{4} =$	$\dfrac{3}{12} =$	$\dfrac{3}{15} =$
$\dfrac{3}{9} =$	$\dfrac{2}{12} =$	$\dfrac{6}{10} =$	$\dfrac{12}{16} =$	$\dfrac{50}{100} =$
$\dfrac{9}{12} =$	$\dfrac{3}{6} =$	$\dfrac{5}{15} =$	$\dfrac{10}{12} =$	$\dfrac{8}{24} =$
$\dfrac{12}{15} =$	$\dfrac{8}{10} =$	$\dfrac{75}{100} =$	$\dfrac{6}{12} =$	$\dfrac{12}{24} =$

I

40 Fractions to Reduce
For use with Lesson 92

Name _____

Time _____

Reduce each fraction to lowest terms.

$\frac{2}{10} =$	$\frac{8}{16} =$	$\frac{2}{6} =$	$\frac{10}{100} =$	$\frac{6}{8} =$
$\frac{10}{15} =$	$\frac{5}{10} =$	$\frac{8}{12} =$	$\frac{9}{15} =$	$\frac{4}{16} =$
$\frac{2}{8} =$	$\frac{4}{10} =$	$\frac{15}{20} =$	$\frac{4}{8} =$	$\frac{4}{6} =$
$\frac{6}{15} =$	$\frac{4}{12} =$	$\frac{25}{100} =$	$\frac{10}{25} =$	$\frac{12}{20} =$
$\frac{20}{100} =$	$\frac{6}{9} =$	$\frac{2}{4} =$	$\frac{3}{12} =$	$\frac{3}{15} =$
$\frac{3}{9} =$	$\frac{2}{12} =$	$\frac{6}{10} =$	$\frac{12}{16} =$	$\frac{50}{100} =$
$\frac{9}{12} =$	$\frac{3}{6} =$	$\frac{5}{15} =$	$\frac{10}{12} =$	$\frac{8}{24} =$
$\frac{12}{15} =$	$\frac{8}{10} =$	$\frac{75}{100} =$	$\frac{6}{12} =$	$\frac{12}{24} =$

I

40 Fractions to Reduce
For use with Lesson 93

Name _____

Time _____

Reduce each fraction to lowest terms.

$\frac{2}{10} =$	$\frac{8}{16} =$	$\frac{2}{6} =$	$\frac{10}{100} =$	$\frac{6}{8} =$
$\frac{10}{15} =$	$\frac{5}{10} =$	$\frac{8}{12} =$	$\frac{9}{15} =$	$\frac{4}{16} =$
$\frac{2}{8} =$	$\frac{4}{10} =$	$\frac{15}{20} =$	$\frac{4}{8} =$	$\frac{4}{6} =$
$\frac{6}{15} =$	$\frac{4}{12} =$	$\frac{25}{100} =$	$\frac{10}{25} =$	$\frac{12}{20} =$
$\frac{20}{100} =$	$\frac{6}{9} =$	$\frac{2}{4} =$	$\frac{3}{12} =$	$\frac{3}{15} =$
$\frac{3}{9} =$	$\frac{2}{12} =$	$\frac{6}{10} =$	$\frac{12}{16} =$	$\frac{50}{100} =$
$\frac{9}{12} =$	$\frac{3}{6} =$	$\frac{5}{15} =$	$\frac{10}{12} =$	$\frac{8}{24} =$
$\frac{12}{15} =$	$\frac{8}{10} =$	$\frac{75}{100} =$	$\frac{6}{12} =$	$\frac{12}{24} =$

I

40 Fractions to Reduce
For use with Lesson 94

Name _____

Time _____

Reduce each fraction to lowest terms.

$\frac{2}{10} =$	$\frac{8}{16} =$	$\frac{2}{6} =$	$\frac{10}{100} =$	$\frac{6}{8} =$
$\frac{10}{15} =$	$\frac{5}{10} =$	$\frac{8}{12} =$	$\frac{9}{15} =$	$\frac{4}{16} =$
$\frac{2}{8} =$	$\frac{4}{10} =$	$\frac{15}{20} =$	$\frac{4}{8} =$	$\frac{4}{6} =$
$\frac{6}{15} =$	$\frac{4}{12} =$	$\frac{25}{100} =$	$\frac{10}{25} =$	$\frac{12}{20} =$
$\frac{20}{100} =$	$\frac{6}{9} =$	$\frac{2}{4} =$	$\frac{3}{12} =$	$\frac{3}{15} =$
$\frac{3}{9} =$	$\frac{2}{12} =$	$\frac{6}{10} =$	$\frac{12}{16} =$	$\frac{50}{100} =$
$\frac{9}{12} =$	$\frac{3}{6} =$	$\frac{5}{15} =$	$\frac{10}{12} =$	$\frac{8}{24} =$
$\frac{12}{15} =$	$\frac{8}{10} =$	$\frac{75}{100} =$	$\frac{6}{12} =$	$\frac{12}{24} =$

I	**40 Fractions to Reduce**	Name _____
	For use with Lesson 95	Time _____

Reduce each fraction to lowest terms.

$\frac{2}{10} =$	$\frac{8}{16} =$	$\frac{2}{6} =$	$\frac{10}{100} =$	$\frac{6}{8} =$
$\frac{10}{15} =$	$\frac{5}{10} =$	$\frac{8}{12} =$	$\frac{9}{15} =$	$\frac{4}{16} =$
$\frac{2}{8} =$	$\frac{4}{10} =$	$\frac{15}{20} =$	$\frac{4}{8} =$	$\frac{4}{6} =$
$\frac{6}{15} =$	$\frac{4}{12} =$	$\frac{25}{100} =$	$\frac{10}{25} =$	$\frac{12}{20} =$
$\frac{20}{100} =$	$\frac{6}{9} =$	$\frac{2}{4} =$	$\frac{3}{12} =$	$\frac{3}{15} =$
$\frac{3}{9} =$	$\frac{2}{12} =$	$\frac{6}{10} =$	$\frac{12}{16} =$	$\frac{50}{100} =$
$\frac{9}{12} =$	$\frac{3}{6} =$	$\frac{5}{15} =$	$\frac{10}{12} =$	$\frac{8}{24} =$
$\frac{12}{15} =$	$\frac{8}{10} =$	$\frac{75}{100} =$	$\frac{6}{12} =$	$\frac{12}{24} =$

I

40 Fractions to Reduce
For use with Test 18

Name _____

Time _____

Reduce each fraction to lowest terms.

$\frac{2}{10} =$	$\frac{8}{16} =$	$\frac{2}{6} =$	$\frac{10}{100} =$	$\frac{6}{8} =$
$\frac{10}{15} =$	$\frac{5}{10} =$	$\frac{8}{12} =$	$\frac{9}{15} =$	$\frac{4}{16} =$
$\frac{2}{8} =$	$\frac{4}{10} =$	$\frac{15}{20} =$	$\frac{4}{8} =$	$\frac{4}{6} =$
$\frac{6}{15} =$	$\frac{4}{12} =$	$\frac{25}{100} =$	$\frac{10}{25} =$	$\frac{12}{20} =$
$\frac{20}{100} =$	$\frac{6}{9} =$	$\frac{2}{4} =$	$\frac{3}{12} =$	$\frac{3}{15} =$
$\frac{3}{9} =$	$\frac{2}{12} =$	$\frac{6}{10} =$	$\frac{12}{16} =$	$\frac{50}{100} =$
$\frac{9}{12} =$	$\frac{3}{6} =$	$\frac{5}{15} =$	$\frac{10}{12} =$	$\frac{8}{24} =$
$\frac{12}{15} =$	$\frac{8}{10} =$	$\frac{75}{100} =$	$\frac{6}{12} =$	$\frac{12}{24} =$

I

40 Fractions to Reduce
For use with Lesson 96

Name _____

Time _____

Reduce each fraction to lowest terms.

$\frac{2}{10} =$	$\frac{8}{16} =$	$\frac{2}{6} =$	$\frac{10}{100} =$	$\frac{6}{8} =$
$\frac{10}{15} =$	$\frac{5}{10} =$	$\frac{8}{12} =$	$\frac{9}{15} =$	$\frac{4}{16} =$
$\frac{2}{8} =$	$\frac{4}{10} =$	$\frac{15}{20} =$	$\frac{4}{8} =$	$\frac{4}{6} =$
$\frac{6}{15} =$	$\frac{4}{12} =$	$\frac{25}{100} =$	$\frac{10}{25} =$	$\frac{12}{20} =$
$\frac{20}{100} =$	$\frac{6}{9} =$	$\frac{2}{4} =$	$\frac{3}{12} =$	$\frac{3}{15} =$
$\frac{3}{9} =$	$\frac{2}{12} =$	$\frac{6}{10} =$	$\frac{12}{16} =$	$\frac{50}{100} =$
$\frac{9}{12} =$	$\frac{3}{6} =$	$\frac{5}{15} =$	$\frac{10}{12} =$	$\frac{8}{24} =$
$\frac{12}{15} =$	$\frac{8}{10} =$	$\frac{75}{100} =$	$\frac{6}{12} =$	$\frac{12}{24} =$

I

40 Fractions to Reduce
For use with Lesson 97

Name _____

Time _____

Reduce each fraction to lowest terms.

$\frac{2}{10} =$	$\frac{8}{16} =$	$\frac{2}{6} =$	$\frac{10}{100} =$	$\frac{6}{8} =$
$\frac{10}{15} =$	$\frac{5}{10} =$	$\frac{8}{12} =$	$\frac{9}{15} =$	$\frac{4}{16} =$
$\frac{2}{8} =$	$\frac{4}{10} =$	$\frac{15}{20} =$	$\frac{4}{8} =$	$\frac{4}{6} =$
$\frac{6}{15} =$	$\frac{4}{12} =$	$\frac{25}{100} =$	$\frac{10}{25} =$	$\frac{12}{20} =$
$\frac{20}{100} =$	$\frac{6}{9} =$	$\frac{2}{4} =$	$\frac{3}{12} =$	$\frac{3}{15} =$
$\frac{3}{9} =$	$\frac{2}{12} =$	$\frac{6}{10} =$	$\frac{12}{16} =$	$\frac{50}{100} =$
$\frac{9}{12} =$	$\frac{3}{6} =$	$\frac{5}{15} =$	$\frac{10}{12} =$	$\frac{8}{24} =$
$\frac{12}{15} =$	$\frac{8}{10} =$	$\frac{75}{100} =$	$\frac{6}{12} =$	$\frac{12}{24} =$

I | 40 Fractions to Reduce
For use with Lesson 98

Name _____

Time _____

Reduce each fraction to lowest terms.

$\dfrac{2}{10} =$	$\dfrac{8}{16} =$	$\dfrac{2}{6} =$	$\dfrac{10}{100} =$	$\dfrac{6}{8} =$
$\dfrac{10}{15} =$	$\dfrac{5}{10} =$	$\dfrac{8}{12} =$	$\dfrac{9}{15} =$	$\dfrac{4}{16} =$
$\dfrac{2}{8} =$	$\dfrac{4}{10} =$	$\dfrac{15}{20} =$	$\dfrac{4}{8} =$	$\dfrac{4}{6} =$
$\dfrac{6}{15} =$	$\dfrac{4}{12} =$	$\dfrac{25}{100} =$	$\dfrac{10}{25} =$	$\dfrac{12}{20} =$
$\dfrac{20}{100} =$	$\dfrac{6}{9} =$	$\dfrac{2}{4} =$	$\dfrac{3}{12} =$	$\dfrac{3}{15} =$
$\dfrac{3}{9} =$	$\dfrac{2}{12} =$	$\dfrac{6}{10} =$	$\dfrac{12}{16} =$	$\dfrac{50}{100} =$
$\dfrac{9}{12} =$	$\dfrac{3}{6} =$	$\dfrac{5}{15} =$	$\dfrac{10}{12} =$	$\dfrac{8}{24} =$
$\dfrac{12}{15} =$	$\dfrac{8}{10} =$	$\dfrac{75}{100} =$	$\dfrac{6}{12} =$	$\dfrac{12}{24} =$

I

40 Fractions to Reduce
For use with Lesson 99

Name _____

Time _____

Reduce each fraction to lowest terms.

$\frac{2}{10} =$	$\frac{8}{16} =$	$\frac{2}{6} =$	$\frac{10}{100} =$	$\frac{6}{8} =$
$\frac{10}{15} =$	$\frac{5}{10} =$	$\frac{8}{12} =$	$\frac{9}{15} =$	$\frac{4}{16} =$
$\frac{2}{8} =$	$\frac{4}{10} =$	$\frac{15}{20} =$	$\frac{4}{8} =$	$\frac{4}{6} =$
$\frac{6}{15} =$	$\frac{4}{12} =$	$\frac{25}{100} =$	$\frac{10}{25} =$	$\frac{12}{20} =$
$\frac{20}{100} =$	$\frac{6}{9} =$	$\frac{2}{4} =$	$\frac{3}{12} =$	$\frac{3}{15} =$
$\frac{3}{9} =$	$\frac{2}{12} =$	$\frac{6}{10} =$	$\frac{12}{16} =$	$\frac{50}{100} =$
$\frac{9}{12} =$	$\frac{3}{6} =$	$\frac{5}{15} =$	$\frac{10}{12} =$	$\frac{8}{24} =$
$\frac{12}{15} =$	$\frac{8}{10} =$	$\frac{75}{100} =$	$\frac{6}{12} =$	$\frac{12}{24} =$

Saxon Math 6/5—Homeschool

I | 40 Fractions to Reduce

For use with Lesson 100

Name _____

Time _____

Reduce each fraction to lowest terms.

$\frac{2}{10} =$	$\frac{8}{16} =$	$\frac{2}{6} =$	$\frac{10}{100} =$	$\frac{6}{8} =$
$\frac{10}{15} =$	$\frac{5}{10} =$	$\frac{8}{12} =$	$\frac{9}{15} =$	$\frac{4}{16} =$
$\frac{2}{8} =$	$\frac{4}{10} =$	$\frac{15}{20} =$	$\frac{4}{8} =$	$\frac{4}{6} =$
$\frac{6}{15} =$	$\frac{4}{12} =$	$\frac{25}{100} =$	$\frac{10}{25} =$	$\frac{12}{20} =$
$\frac{20}{100} =$	$\frac{6}{9} =$	$\frac{2}{4} =$	$\frac{3}{12} =$	$\frac{3}{15} =$
$\frac{3}{9} =$	$\frac{2}{12} =$	$\frac{6}{10} =$	$\frac{12}{16} =$	$\frac{50}{100} =$
$\frac{9}{12} =$	$\frac{3}{6} =$	$\frac{5}{15} =$	$\frac{10}{12} =$	$\frac{8}{24} =$
$\frac{12}{15} =$	$\frac{8}{10} =$	$\frac{75}{100} =$	$\frac{6}{12} =$	$\frac{12}{24} =$

I

40 Fractions to Reduce
For use with Test 19

Name _____

Time _____

Reduce each fraction to lowest terms.

$\frac{2}{10} =$	$\frac{8}{16} =$	$\frac{2}{6} =$	$\frac{10}{100} =$	$\frac{6}{8} =$
$\frac{10}{15} =$	$\frac{5}{10} =$	$\frac{8}{12} =$	$\frac{9}{15} =$	$\frac{4}{16} =$
$\frac{2}{8} =$	$\frac{4}{10} =$	$\frac{15}{20} =$	$\frac{4}{8} =$	$\frac{4}{6} =$
$\frac{6}{15} =$	$\frac{4}{12} =$	$\frac{25}{100} =$	$\frac{10}{25} =$	$\frac{12}{20} =$
$\frac{20}{100} =$	$\frac{6}{9} =$	$\frac{2}{4} =$	$\frac{3}{12} =$	$\frac{3}{15} =$
$\frac{3}{9} =$	$\frac{2}{12} =$	$\frac{6}{10} =$	$\frac{12}{16} =$	$\frac{50}{100} =$
$\frac{9}{12} =$	$\frac{3}{6} =$	$\frac{5}{15} =$	$\frac{10}{12} =$	$\frac{8}{24} =$
$\frac{12}{15} =$	$\frac{8}{10} =$	$\frac{75}{100} =$	$\frac{6}{12} =$	$\frac{12}{24} =$

Saxon Math 6/5—Homeschool

25

Coordinate Plane
For use with Investigation 10

Name _____

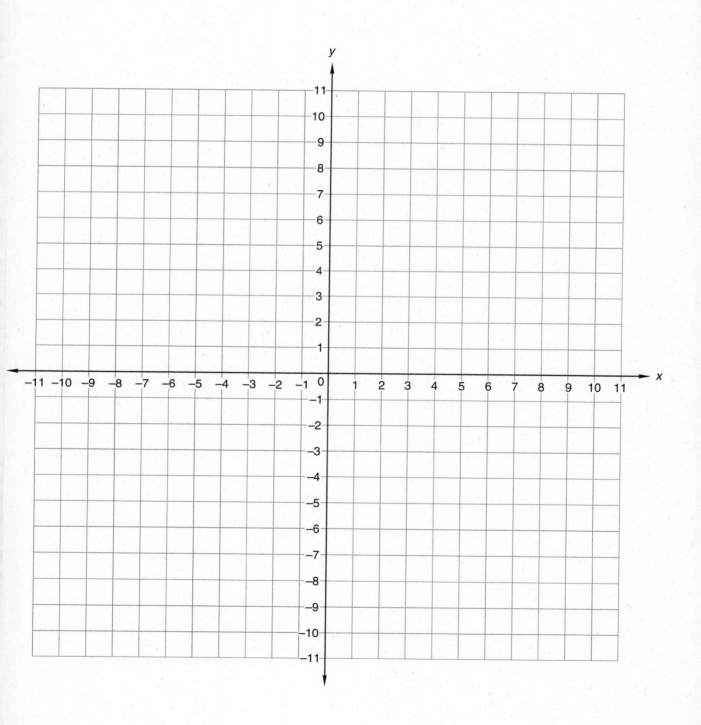

26 | Coordinate Plane

For use with Investigation 10

Name _____

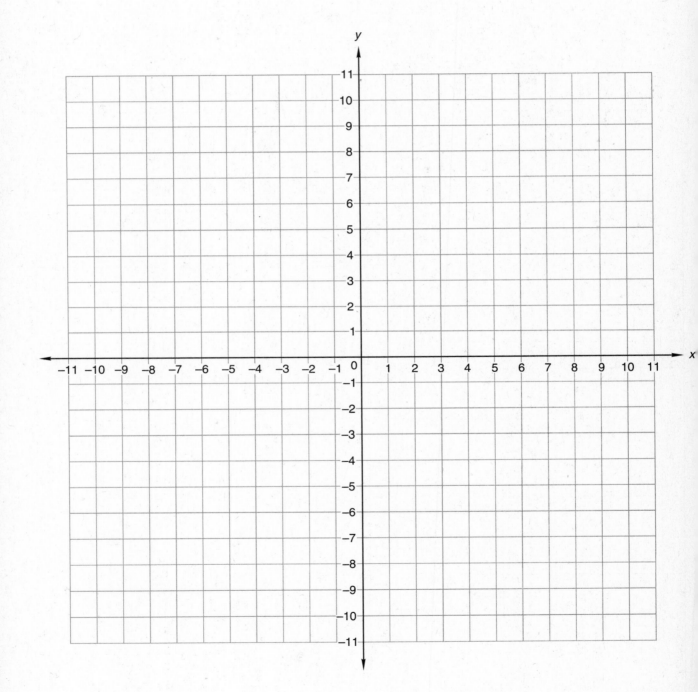

50 Fractions to Simplify

For use with Lesson 101

Name _____

Time _____

Simplify.

$\dfrac{16}{20} =$	$\dfrac{6}{4} =$	$\dfrac{4}{6} =$	$\dfrac{10}{8} =$	$\dfrac{3}{12} =$
$\dfrac{12}{9} =$	$\dfrac{2}{4} =$	$\dfrac{12}{10} =$	$\dfrac{12}{4} =$	$\dfrac{12}{8} =$
$\dfrac{8}{3} =$	$\dfrac{8}{6} =$	$\dfrac{4}{12} =$	$\dfrac{10}{4} =$	$\dfrac{4}{10} =$
$\dfrac{20}{8} =$	$\dfrac{4}{8} =$	$\dfrac{20}{9} =$	$\dfrac{24}{6} =$	$\dfrac{9}{6} =$
$\dfrac{15}{10} =$	$\dfrac{5}{2} =$	$\dfrac{12}{20} =$	$\dfrac{15}{9} =$	$\dfrac{8}{12} =$
$\dfrac{4}{20} =$	$\dfrac{8}{24} =$	$\dfrac{10}{6} =$	$\dfrac{3}{6} =$	$\dfrac{16}{10} =$
$\dfrac{2}{8} =$	$\dfrac{20}{6} =$	$\dfrac{6}{3} =$	$\dfrac{25}{12} =$	$\dfrac{9}{12} =$
$\dfrac{10}{2} =$	$\dfrac{8}{8} =$	$\dfrac{50}{100} =$	$\dfrac{6}{12} =$	$\dfrac{15}{6} =$
$\dfrac{10}{3} =$	$\dfrac{10}{20} =$	$\dfrac{24}{9} =$	$\dfrac{6}{8} =$	$\dfrac{16}{5} =$
$\dfrac{5}{10} =$	$\dfrac{14}{8} =$	$\dfrac{15}{2} =$	$\dfrac{21}{6} =$	$\dfrac{16}{24} =$

J 50 Fractions to Simplify

For use with Lesson 102

Name _____

Time _____

Simplify.

$\frac{16}{20} =$	$\frac{6}{4} =$	$\frac{4}{6} =$	$\frac{10}{8} =$	$\frac{3}{12} =$
$\frac{12}{9} =$	$\frac{2}{4} =$	$\frac{12}{10} =$	$\frac{12}{4} =$	$\frac{12}{8} =$
$\frac{8}{3} =$	$\frac{8}{6} =$	$\frac{4}{12} =$	$\frac{10}{4} =$	$\frac{4}{10} =$
$\frac{20}{8} =$	$\frac{4}{8} =$	$\frac{20}{9} =$	$\frac{24}{6} =$	$\frac{9}{6} =$
$\frac{15}{10} =$	$\frac{5}{2} =$	$\frac{12}{20} =$	$\frac{15}{9} =$	$\frac{8}{12} =$
$\frac{4}{20} =$	$\frac{8}{24} =$	$\frac{10}{6} =$	$\frac{3}{6} =$	$\frac{16}{10} =$
$\frac{2}{8} =$	$\frac{20}{6} =$	$\frac{6}{3} =$	$\frac{25}{12} =$	$\frac{9}{12} =$
$\frac{10}{2} =$	$\frac{8}{8} =$	$\frac{50}{100} =$	$\frac{6}{12} =$	$\frac{15}{6} =$
$\frac{10}{3} =$	$\frac{10}{20} =$	$\frac{24}{9} =$	$\frac{6}{8} =$	$\frac{16}{5} =$
$\frac{5}{10} =$	$\frac{14}{8} =$	$\frac{15}{2} =$	$\frac{21}{6} =$	$\frac{16}{24} =$

50 Fractions to Simplify

For use with Lesson 103

Name _____

Time _____

Simplify.

$\frac{16}{20} =$	$\frac{6}{4} =$	$\frac{4}{6} =$	$\frac{10}{8} =$	$\frac{3}{12} =$
$\frac{12}{9} =$	$\frac{2}{4} =$	$\frac{12}{10} =$	$\frac{12}{4} =$	$\frac{12}{8} =$
$\frac{8}{3} =$	$\frac{8}{6} =$	$\frac{4}{12} =$	$\frac{10}{4} =$	$\frac{4}{10} =$
$\frac{20}{8} =$	$\frac{4}{8} =$	$\frac{20}{9} =$	$\frac{24}{6} =$	$\frac{9}{6} =$
$\frac{15}{10} =$	$\frac{5}{2} =$	$\frac{12}{20} =$	$\frac{15}{9} =$	$\frac{8}{12} =$
$\frac{4}{20} =$	$\frac{8}{24} =$	$\frac{10}{6} =$	$\frac{3}{6} =$	$\frac{16}{10} =$
$\frac{2}{8} =$	$\frac{20}{6} =$	$\frac{6}{3} =$	$\frac{25}{12} =$	$\frac{9}{12} =$
$\frac{10}{2} =$	$\frac{8}{8} =$	$\frac{50}{100} =$	$\frac{6}{12} =$	$\frac{15}{6} =$
$\frac{10}{3} =$	$\frac{10}{20} =$	$\frac{24}{9} =$	$\frac{6}{8} =$	$\frac{16}{5} =$
$\frac{5}{10} =$	$\frac{14}{8} =$	$\frac{15}{2} =$	$\frac{21}{6} =$	$\frac{16}{24} =$

50 Fractions to Simplify
For use with Lesson 104

Name _____

Time _____

Simplify.

$\frac{16}{20} =$	$\frac{6}{4} =$	$\frac{4}{6} =$	$\frac{10}{8} =$	$\frac{3}{12} =$
$\frac{12}{9} =$	$\frac{2}{4} =$	$\frac{12}{10} =$	$\frac{12}{4} =$	$\frac{12}{8} =$
$\frac{8}{3} =$	$\frac{8}{6} =$	$\frac{4}{12} =$	$\frac{10}{4} =$	$\frac{4}{10} =$
$\frac{20}{8} =$	$\frac{4}{8} =$	$\frac{20}{9} =$	$\frac{24}{6} =$	$\frac{9}{6} =$
$\frac{15}{10} =$	$\frac{5}{2} =$	$\frac{12}{20} =$	$\frac{15}{9} =$	$\frac{8}{12} =$
$\frac{4}{20} =$	$\frac{8}{24} =$	$\frac{10}{6} =$	$\frac{3}{6} =$	$\frac{16}{10} =$
$\frac{2}{8} =$	$\frac{20}{6} =$	$\frac{6}{3} =$	$\frac{25}{12} =$	$\frac{9}{12} =$
$\frac{10}{2} =$	$\frac{8}{8} =$	$\frac{50}{100} =$	$\frac{6}{12} =$	$\frac{15}{6} =$
$\frac{10}{3} =$	$\frac{10}{20} =$	$\frac{24}{9} =$	$\frac{6}{8} =$	$\frac{16}{5} =$
$\frac{5}{10} =$	$\frac{14}{8} =$	$\frac{15}{2} =$	$\frac{21}{6} =$	$\frac{16}{24} =$

50 Fractions to Simplify
For use with Lesson 105

Name _____

Time _____

Simplify.

$\frac{16}{20} =$	$\frac{6}{4} =$	$\frac{4}{6} =$	$\frac{10}{8} =$	$\frac{3}{12} =$
$\frac{12}{9} =$	$\frac{2}{4} =$	$\frac{12}{10} =$	$\frac{12}{4} =$	$\frac{12}{8} =$
$\frac{8}{3} =$	$\frac{8}{6} =$	$\frac{4}{12} =$	$\frac{10}{4} =$	$\frac{4}{10} =$
$\frac{20}{8} =$	$\frac{4}{8} =$	$\frac{20}{9} =$	$\frac{24}{6} =$	$\frac{9}{6} =$
$\frac{15}{10} =$	$\frac{5}{2} =$	$\frac{12}{20} =$	$\frac{15}{9} =$	$\frac{8}{12} =$
$\frac{4}{20} =$	$\frac{8}{24} =$	$\frac{10}{6} =$	$\frac{3}{6} =$	$\frac{16}{10} =$
$\frac{2}{8} =$	$\frac{20}{6} =$	$\frac{6}{3} =$	$\frac{25}{12} =$	$\frac{9}{12} =$
$\frac{10}{2} =$	$\frac{8}{8} =$	$\frac{50}{100} =$	$\frac{6}{12} =$	$\frac{15}{6} =$
$\frac{10}{3} =$	$\frac{10}{20} =$	$\frac{24}{9} =$	$\frac{6}{8} =$	$\frac{16}{5} =$
$\frac{5}{10} =$	$\frac{14}{8} =$	$\frac{15}{2} =$	$\frac{21}{6} =$	$\frac{16}{24} =$

J

50 Fractions to Simplify
For use with Test 20

Name _____

Time _____

Simplify.

$\frac{16}{20} =$	$\frac{6}{4} =$	$\frac{4}{6} =$	$\frac{10}{8} =$	$\frac{3}{12} =$
$\frac{12}{9} =$	$\frac{2}{4} =$	$\frac{12}{10} =$	$\frac{12}{4} =$	$\frac{12}{8} =$
$\frac{8}{3} =$	$\frac{8}{6} =$	$\frac{4}{12} =$	$\frac{10}{4} =$	$\frac{4}{10} =$
$\frac{20}{8} =$	$\frac{4}{8} =$	$\frac{20}{9} =$	$\frac{24}{6} =$	$\frac{9}{6} =$
$\frac{15}{10} =$	$\frac{5}{2} =$	$\frac{12}{20} =$	$\frac{15}{9} =$	$\frac{8}{12} =$
$\frac{4}{20} =$	$\frac{8}{24} =$	$\frac{10}{6} =$	$\frac{3}{6} =$	$\frac{16}{10} =$
$\frac{2}{8} =$	$\frac{20}{6} =$	$\frac{6}{3} =$	$\frac{25}{12} =$	$\frac{9}{12} =$
$\frac{10}{2} =$	$\frac{8}{8} =$	$\frac{50}{100} =$	$\frac{6}{12} =$	$\frac{15}{6} =$
$\frac{10}{3} =$	$\frac{10}{20} =$	$\frac{24}{9} =$	$\frac{6}{8} =$	$\frac{16}{5} =$
$\frac{5}{10} =$	$\frac{14}{8} =$	$\frac{15}{2} =$	$\frac{21}{6} =$	$\frac{16}{24} =$

J

50 Fractions to Simplify
For use with Lesson 106

Name _____

Time _____

Simplify.

$\frac{16}{20} =$	$\frac{6}{4} =$	$\frac{4}{6} =$	$\frac{10}{8} =$	$\frac{3}{12} =$
$\frac{12}{9} =$	$\frac{2}{4} =$	$\frac{12}{10} =$	$\frac{12}{4} =$	$\frac{12}{8} =$
$\frac{8}{3} =$	$\frac{8}{6} =$	$\frac{4}{12} =$	$\frac{10}{4} =$	$\frac{4}{10} =$
$\frac{20}{8} =$	$\frac{4}{8} =$	$\frac{20}{9} =$	$\frac{24}{6} =$	$\frac{9}{6} =$
$\frac{15}{10} =$	$\frac{5}{2} =$	$\frac{12}{20} =$	$\frac{15}{9} =$	$\frac{8}{12} =$
$\frac{4}{20} =$	$\frac{8}{24} =$	$\frac{10}{6} =$	$\frac{3}{6} =$	$\frac{16}{10} =$
$\frac{2}{8} =$	$\frac{20}{6} =$	$\frac{6}{3} =$	$\frac{25}{12} =$	$\frac{9}{12} =$
$\frac{10}{2} =$	$\frac{8}{8} =$	$\frac{50}{100} =$	$\frac{6}{12} =$	$\frac{15}{6} =$
$\frac{10}{3} =$	$\frac{10}{20} =$	$\frac{24}{9} =$	$\frac{6}{8} =$	$\frac{16}{5} =$
$\frac{5}{10} =$	$\frac{14}{8} =$	$\frac{15}{2} =$	$\frac{21}{6} =$	$\frac{16}{24} =$

Saxon Math 6/5—Homeschool

J | 50 Fractions to Simplify

For use with Lesson 107

Name _____

Time _____

Simplify.

$\frac{16}{20} =$	$\frac{6}{4} =$	$\frac{4}{6} =$	$\frac{10}{8} =$	$\frac{3}{12} =$
$\frac{12}{9} =$	$\frac{2}{4} =$	$\frac{12}{10} =$	$\frac{12}{4} =$	$\frac{12}{8} =$
$\frac{8}{3} =$	$\frac{8}{6} =$	$\frac{4}{12} =$	$\frac{10}{4} =$	$\frac{4}{10} =$
$\frac{20}{8} =$	$\frac{4}{8} =$	$\frac{20}{9} =$	$\frac{24}{6} =$	$\frac{9}{6} =$
$\frac{15}{10} =$	$\frac{5}{2} =$	$\frac{12}{20} =$	$\frac{15}{9} =$	$\frac{8}{12} =$
$\frac{4}{20} =$	$\frac{8}{24} =$	$\frac{10}{6} =$	$\frac{3}{6} =$	$\frac{16}{10} =$
$\frac{2}{8} =$	$\frac{20}{6} =$	$\frac{6}{3} =$	$\frac{25}{12} =$	$\frac{9}{12} =$
$\frac{10}{2} =$	$\frac{8}{8} =$	$\frac{50}{100} =$	$\frac{6}{12} =$	$\frac{15}{6} =$
$\frac{10}{3} =$	$\frac{10}{20} =$	$\frac{24}{9} =$	$\frac{6}{8} =$	$\frac{16}{5} =$
$\frac{5}{10} =$	$\frac{14}{8} =$	$\frac{15}{2} =$	$\frac{21}{6} =$	$\frac{16}{24} =$

Saxon Math 6/5—Homeschool

50 Fractions to Simplify
For use with Lesson 108

Name _____

Time _____

Simplify.

$\frac{16}{20} =$	$\frac{6}{4} =$	$\frac{4}{6} =$	$\frac{10}{8} =$	$\frac{3}{12} =$
$\frac{12}{9} =$	$\frac{2}{4} =$	$\frac{12}{10} =$	$\frac{12}{4} =$	$\frac{12}{8} =$
$\frac{8}{3} =$	$\frac{8}{6} =$	$\frac{4}{12} =$	$\frac{10}{4} =$	$\frac{4}{10} =$
$\frac{20}{8} =$	$\frac{4}{8} =$	$\frac{20}{9} =$	$\frac{24}{6} =$	$\frac{9}{6} =$
$\frac{15}{10} =$	$\frac{5}{2} =$	$\frac{12}{20} =$	$\frac{15}{9} =$	$\frac{8}{12} =$
$\frac{4}{20} =$	$\frac{8}{24} =$	$\frac{10}{6} =$	$\frac{3}{6} =$	$\frac{16}{10} =$
$\frac{2}{8} =$	$\frac{20}{6} =$	$\frac{6}{3} =$	$\frac{25}{12} =$	$\frac{9}{12} =$
$\frac{10}{2} =$	$\frac{8}{8} =$	$\frac{50}{100} =$	$\frac{6}{12} =$	$\frac{15}{6} =$
$\frac{10}{3} =$	$\frac{10}{20} =$	$\frac{24}{9} =$	$\frac{6}{8} =$	$\frac{16}{5} =$
$\frac{5}{10} =$	$\frac{14}{8} =$	$\frac{15}{2} =$	$\frac{21}{6} =$	$\frac{16}{24} =$

50 Fractions to Simplify
For use with Lesson 109

Name _____

Time _____

Simplify.

$\frac{16}{20} =$	$\frac{6}{4} =$	$\frac{4}{6} =$	$\frac{10}{8} =$	$\frac{3}{12} =$
$\frac{12}{9} =$	$\frac{2}{4} =$	$\frac{12}{10} =$	$\frac{12}{4} =$	$\frac{12}{8} =$
$\frac{8}{3} =$	$\frac{8}{6} =$	$\frac{4}{12} =$	$\frac{10}{4} =$	$\frac{4}{10} =$
$\frac{20}{8} =$	$\frac{4}{8} =$	$\frac{20}{9} =$	$\frac{24}{6} =$	$\frac{9}{6} =$
$\frac{15}{10} =$	$\frac{5}{2} =$	$\frac{12}{20} =$	$\frac{15}{9} =$	$\frac{8}{12} =$
$\frac{4}{20} =$	$\frac{8}{24} =$	$\frac{10}{6} =$	$\frac{3}{6} =$	$\frac{16}{10} =$
$\frac{2}{8} =$	$\frac{20}{6} =$	$\frac{6}{3} =$	$\frac{25}{12} =$	$\frac{9}{12} =$
$\frac{10}{2} =$	$\frac{8}{8} =$	$\frac{50}{100} =$	$\frac{6}{12} =$	$\frac{15}{6} =$
$\frac{10}{3} =$	$\frac{10}{20} =$	$\frac{24}{9} =$	$\frac{6}{8} =$	$\frac{16}{5} =$
$\frac{5}{10} =$	$\frac{14}{8} =$	$\frac{15}{2} =$	$\frac{21}{6} =$	$\frac{16}{24} =$

FACTS PRACTICE TEST

50 Fractions to Simplify
For use with Lesson 110

Name _____

Time _____

Simplify.

$\frac{16}{20} =$	$\frac{6}{4} =$	$\frac{4}{6} =$	$\frac{10}{8} =$	$\frac{3}{12} =$
$\frac{12}{9} =$	$\frac{2}{4} =$	$\frac{12}{10} =$	$\frac{12}{4} =$	$\frac{12}{8} =$
$\frac{8}{3} =$	$\frac{8}{6} =$	$\frac{4}{12} =$	$\frac{10}{4} =$	$\frac{4}{10} =$
$\frac{20}{8} =$	$\frac{4}{8} =$	$\frac{20}{9} =$	$\frac{24}{6} =$	$\frac{9}{6} =$
$\frac{15}{10} =$	$\frac{5}{2} =$	$\frac{12}{20} =$	$\frac{15}{9} =$	$\frac{8}{12} =$
$\frac{4}{20} =$	$\frac{8}{24} =$	$\frac{10}{6} =$	$\frac{3}{6} =$	$\frac{16}{10} =$
$\frac{2}{8} =$	$\frac{20}{6} =$	$\frac{6}{3} =$	$\frac{25}{12} =$	$\frac{9}{12} =$
$\frac{10}{2} =$	$\frac{8}{8} =$	$\frac{50}{100} =$	$\frac{6}{12} =$	$\frac{15}{6} =$
$\frac{10}{3} =$	$\frac{10}{20} =$	$\frac{24}{9} =$	$\frac{6}{8} =$	$\frac{16}{5} =$
$\frac{5}{10} =$	$\frac{14}{8} =$	$\frac{15}{2} =$	$\frac{21}{6} =$	$\frac{16}{24} =$

Saxon Math 6/5—Homeschool

50 Fractions to Simplify
For use with Test 21

Name _____

Time _____

Simplify.

$\frac{16}{20} =$	$\frac{6}{4} =$	$\frac{4}{6} =$	$\frac{10}{8} =$	$\frac{3}{12} =$
$\frac{12}{9} =$	$\frac{2}{4} =$	$\frac{12}{10} =$	$\frac{12}{4} =$	$\frac{12}{8} =$
$\frac{8}{3} =$	$\frac{8}{6} =$	$\frac{4}{12} =$	$\frac{10}{4} =$	$\frac{4}{10} =$
$\frac{20}{8} =$	$\frac{4}{8} =$	$\frac{20}{9} =$	$\frac{24}{6} =$	$\frac{9}{6} =$
$\frac{15}{10} =$	$\frac{5}{2} =$	$\frac{12}{20} =$	$\frac{15}{9} =$	$\frac{8}{12} =$
$\frac{4}{20} =$	$\frac{8}{24} =$	$\frac{10}{6} =$	$\frac{3}{6} =$	$\frac{16}{10} =$
$\frac{2}{8} =$	$\frac{20}{6} =$	$\frac{6}{3} =$	$\frac{25}{12} =$	$\frac{9}{12} =$
$\frac{10}{2} =$	$\frac{8}{8} =$	$\frac{50}{100} =$	$\frac{6}{12} =$	$\frac{15}{6} =$
$\frac{10}{3} =$	$\frac{10}{20} =$	$\frac{24}{9} =$	$\frac{6}{8} =$	$\frac{16}{5} =$
$\frac{5}{10} =$	$\frac{14}{8} =$	$\frac{15}{2} =$	$\frac{21}{6} =$	$\frac{16}{24} =$

K

30 Percents to Write as Fractions

For use with Lesson 111

Name _____

Time _____

Write each percent as a reduced fraction.

1% =	20% =	55% =	90% =	75% =
99% =	5% =	95% =	80% =	12% =
70% =	65% =	50% =	2% =	48% =
24% =	25% =	98% =	40% =	15% =
60% =	30% =	4% =	35% =	36% =
45% =	8% =	10% =	21% =	85% =

K

30 Percents to Write as Fractions
For use with Lesson 112

Name _____

Time _____

Write each percent as a reduced fraction.

1% =	20% =	55% =	90% =	75% =
99% =	5% =	95% =	80% =	12% =
70% =	65% =	50% =	2% =	48% =
24% =	25% =	98% =	40% =	15% =
60% =	30% =	4% =	35% =	36% =
45% =	8% =	10% =	21% =	85% =

30 Percents to Write as Fractions
For use with Lesson 113

Name _____

Time _____

Write each percent as a reduced fraction.

1% =	20% =	55% =	90% =	75% =
99% =	5% =	95% =	80% =	12% =
70% =	65% =	50% =	2% =	48% =
24% =	25% =	98% =	40% =	15% =
60% =	30% =	4% =	35% =	36% =
45% =	8% =	10% =	21% =	85% =

K	**30 Percents to Write as Fractions** *For use with Lesson 114*	Name _____ Time _____

Write each percent as a reduced fraction.

1% =	20% =	55% =	90% =	75% =
99% =	5% =	95% =	80% =	12% =
70% =	65% =	50% =	2% =	48% =
24% =	25% =	98% =	40% =	15% =
60% =	30% =	4% =	35% =	36% =
45% =	8% =	10% =	21% =	85% =

30 Percents to Write as Fractions
For use with Lesson 115

Name _____

Time _____

Write each percent as a reduced fraction.

1% =	20% =	55% =	90% =	75% =
99% =	5% =	95% =	80% =	12% =
70% =	65% =	50% =	2% =	48% =
24% =	25% =	98% =	40% =	15% =
60% =	30% =	4% =	35% =	36% =
45% =	8% =	10% =	21% =	85% =

K | 30 Percents to Write as Fractions
For use with Test 22

Name _____

Time _____

Write each percent as a reduced fraction.

1% =	20% =	55% =	90% =	75% =
99% =	5% =	95% =	80% =	12% =
70% =	65% =	50% =	2% =	48% =
24% =	25% =	98% =	40% =	15% =
60% =	30% =	4% =	35% =	36% =
45% =	8% =	10% =	21% =	85% =

K

30 Percents to Write as Fractions

For use with Lesson 116

Name _____

Time _____

Write each percent as a reduced fraction.

1% =	20% =	55% =	90% =	75% =
99% =	5% =	95% =	80% =	12% =
70% =	65% =	50% =	2% =	48% =
24% =	25% =	98% =	40% =	15% =
60% =	30% =	4% =	35% =	36% =
45% =	8% =	10% =	21% =	85% =

K

30 Percents to Write as Fractions
For use with Lesson 117

Name _____

Time _____

Write each percent as a reduced fraction.

1% =	20% =	55% =	90% =	75% =
99% =	5% =	95% =	80% =	12% =
70% =	65% =	50% =	2% =	48% =
24% =	25% =	98% =	40% =	15% =
60% =	30% =	4% =	35% =	36% =
45% =	8% =	10% =	21% =	85% =

Saxon Math 6/5—Homeschool

K

30 Percents to Write as Fractions
For use with Lesson 119

Name _____

Time _____

Write each percent as a reduced fraction.

1% =	20% =	55% =	90% =	75% =
99% =	5% =	95% =	80% =	12% =
70% =	65% =	50% =	2% =	48% =
24% =	25% =	98% =	40% =	15% =
60% =	30% =	4% =	35% =	36% =
45% =	8% =	10% =	21% =	85% =

K

30 Percents to Write as Fractions

For use with Lesson 120

Name _____

Time _____

Write each percent as a reduced fraction.

1% =	20% =	55% =	90% =	75% =
99% =	5% =	95% =	80% =	12% =
70% =	65% =	50% =	2% =	48% =
24% =	25% =	98% =	40% =	15% =
60% =	30% =	4% =	35% =	36% =
45% =	8% =	10% =	21% =	85% =

Saxon Math 6/5—Homeschool

30 Percents to Write as Fractions
For use with Test 23

Name _____

Time _____

Write each percent as a reduced fraction.

1% =	20% =	55% =	90% =	75% =
99% =	5% =	95% =	80% =	12% =
70% =	65% =	50% =	2% =	48% =
24% =	25% =	98% =	40% =	15% =
60% =	30% =	4% =	35% =	36% =
45% =	8% =	10% =	21% =	85% =

27 | **Tessellations**
For use with Investigation 12

Carefully cut out these polygons. Form a tessellation using the triangles. Then form a tessellation using the quadrilaterals.

28 Tessellations

For use with Investigation 12

Name _____

Follow the instructions in Investigation 12 to create a tessellation in the box below.

Tile in this area.

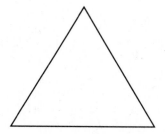

Tests

A test should be given after every fifth lesson, beginning after Lesson 10. The testing schedule is explained in greater detail on the back of this page.

On test days, allow five minutes for your student to take the Facts Practice Test indicated at the top of the test. Then administer the cumulative test specified by the Testing Schedule. You might wish also to provide your student with a photocopy of Recording Form E. This form is designed to provide an organized space for your student to show his or her work. *Note:* The textbook should not be used during the test.

Solutions to the test problems are located in the *Saxon Math 6/5—Homeschool Solutions Manual.* For detailed information on appropriate test-grading strategies, please refer to the preface in the *Saxon Math 6/5—Homeschool* textbook.

Testing Schedule

Test to be administered	Covers material through	Give after
Test 1	Lesson 5	Lesson 10
Test 2	Lesson 10	Lesson 15
Test 3	Lesson 15	Lesson 20
Test 4	Lesson 20	Lesson 25
Test 5	Lesson 25	Lesson 30
Test 6	Lesson 30	Lesson 35
Test 7	Lesson 35	Lesson 40
Test 8	Lesson 40	Lesson 45
Test 9	Lesson 45	Lesson 50
Test 10	Lesson 50	Lesson 55
Test 11	Lesson 55	Lesson 60
Test 12	Lesson 60	Lesson 65
Test 13	Lesson 65	Lesson 70
Test 14	Lesson 70	Lesson 75
Test 15	Lesson 75	Lesson 80
Test 16	Lesson 80	Lesson 85
Test 17	Lesson 85	Lesson 90
Test 18	Lesson 90	Lesson 95
Test 19	Lesson 95	Lesson 100
Test 20	Lesson 100	Lesson 105
Test 21	Lesson 105	Lesson 110
Test 22	Lesson 110	Lesson 115
Test 23	Lesson 115	Lesson 120

*Also take Facts Practice Test A
(100 Addition Facts).*

Name _____

Write the next three terms in each counting sequence:

1. 8, 16, 24, 32, _____, _____, _____, ...
(1)

2. 49, 42, 35, 28, _____, _____, _____, ...
(1)

3. 25, 30, 35, 40, _____, _____, _____, ...
(1)

4. What is the last digit of 27,329?
(1)

5. All whole numbers are either even or what?
(2)

6. Which of these numbers is even?
(2)
    A. 3527   B. 2735   C. 5732

7. Which of these numbers is odd?
(2)
    A. 8431   B. 3418   C. 1834

8. In the drawer there is one more pen than there are pencils. Which of the following could **not** be the total
(2) number of pens and pencils in the drawer?
    A. 3   B. 14   C. 17

9. Use digits to write the number for "2 hundreds plus 4 tens plus 9 ones."
(3)

10. Which digit in 847 shows the number of tens?
(3)

11. How much is 8 hundreds plus 3 tens?
(3)

12. One hundred equals how many tens?
(3)

Write each comparison using digits and a comparison symbol:

13. Thirteen is less than thirty.
(4)

14. Seventeen is greater than six.
(4)

15. Compare: 102 ◯ 111
(4)

16. Compare: 23 ◯ 32
(4)

17. Use words to name $329.72.
(5)

18. Use words to name 115.
(5)

19. Use digits to write seven hundred three dollars and forty cents.
(5)

20. Use digits to write three hundred nine.
(5)

2

Name _____

1. On Pedro's team there are twenty-three girls and eighteen boys. Altogether, how many boys and girls are on his team?
(11)

2. On Siew's team there are 50 boys and girls. Half the athletes are girls. How many girls are on Siew's team?
(2)

3. For the fact family 1, 4, and 5, write two addition facts and two subtraction facts.
(8)

4. Use digits to write three hundred seven thousand, eight hundred thirteen.
(7)

5. Use the three digits 2, 3, and 4 once each to make an even number greater than 400.
(2)

Find each missing addend:

6. $14 + n = 22$
(10)

7. $b + 11 = 50$
(10)

8. Which digit in 82,384 shows the number of hundreds?
(3)

9. Compare: $37 - 7 \bigcirc 32 - 2$
(4, 8)

10. Think of two even numbers. Add them together. Is the sum odd or even?
(2)

11. Kelly is third in line. Bobby is fifteenth in the same line. How many people are between them?
(7)

12. $\begin{array}{r} 143 \\ 87 \\ + 623 \\ \hline \end{array}$
(6)

13. $\begin{array}{r} 327 \\ - 239 \\ \hline \end{array}$
(9)

14. $\begin{array}{r} 900 \\ - 238 \\ \hline \end{array}$
(9)

15. $\begin{array}{r} 5 \\ 6 \\ 2 \\ 9 \\ + 7 \\ \hline \end{array}$
(6)

16. $\$229 - \$40$
(9)

17. $\$23 + \$276 + \$3$
(6)

Write the next term in each counting sequence:

18. 9, 18, 27, _____, …
(1)

19. 35, 40, 45, _____, …
(1)

20. 63, 56, 49, _____, …
(1)

TEST

3

Name _____

1. Write two addition facts and two subtraction facts for the fact family 5, 6, and 11.
(8)

2. What is the product of 4 and 9?
(15)

3. Draw a number line marked with the integers from –8 to 8.
(12)

4. Which figure shows a segment?
(12)

A. ⟵——————⟶   B. ————   C. ——————⟶

5. Find the missing number: $17 - n = 8$
(14)

6. There are 11 boys and 16 girls in the museum. Altogether, how many boys and girls are in the museum?
(11)

7. Camden picked 25 peaches in the morning. If he picked 34 peaches that day, how many peaches did Camden pick in the afternoon?
(11)

8. Use tally marks to show the number fourteen.
(12)

9. Which number sentence illustrates the identity property of multiplication?
(15)

A. $4 \times 1 = 4$ B. $3 \times 4 = 4 \times 3$ C. $4 \times 0 = 0$

10. Change this addition problem to a multiplication problem: $3 + 3 + 3 + 3 + 3 + 3$
(13)

11. $408 - 29$
(9)

12. $\begin{array}{r} \$3.24 \\ \$1.17 \\ + \$7.03 \\ \hline \end{array}$
(13)

13. $\begin{array}{r} 729 \\ 84 \\ + 654 \\ \hline \end{array}$
(6)

14. $\begin{array}{r} \$2.12 \\ - \$1.83 \\ \hline \end{array}$
(13)

15. $\begin{array}{r} 900 \\ - 767 \\ \hline \end{array}$
(9)

16. Which digit is in the tens place in 432,685?
(7)

17. Use digits to write the number forty-seven thousand, nine hundred seventy.
(7)

18. How much is 3 eights?
(15)

19. Compare: $2767 \bigcirc 2776$
(7)

20. What is the **seventh** term in this counting sequence?
(1)

$$8, 16, 24, 32, \ldots$$

4

*Also take Facts Practice Test F
(64 Multiplication Facts).*

Name _____

1. Randy read 2 journals. One journal had 267 pages. The other had 198 pages. How many pages did Randy
(11) read in all?

2. There are four soccer teams with 9 players on each team. How many players are there on all four soccer
(17) teams? Find the answer once by adding and again by multiplying.

3. What is the sum of five hundred fifty and two hundred fifteen?
(6)

4. There are 359 children at the zoo. If 169 children are boys, how many are girls?
(16)

5. Use tally marks to show the number nine.
(12)

6. Compare: −4 ◯ 4
(12)

7. Compare: 8 + 8 + 8 + 8 ◯ 5 × 8
(13)

8. Which of these lines is oblique?
(12)

A.   B. ←——————→   C. ↕

9. Think of an odd number and an even number. Multiply them. Is the product odd or even?
(2, 15)

10. $\begin{array}{r} \$27 \\ \times\quad 6 \\ \hline \end{array}$
(17)

11. $\begin{array}{r} \$4.08 \\ \times\quad 5 \\ \hline \end{array}$
(17)

12. $\begin{array}{r} 4321 \\ -\ 2893 \\ \hline \end{array}$
(9)

13. $\begin{array}{r} \$50.00 \\ -\ \$18.63 \\ \hline \end{array}$
(13)

14. $2 \times 4 \times 6$
(18)

15. $3\overline{)12}$
(20)

16. $12 \div 4$
(20)

17. $\dfrac{12}{6}$
(20)

Find each missing number:

18. $m \times 10 = 90$
(18)

19. $e - 16 = 34$
(14)

20. $426 + m = 714$
(10)

5

Also take Facts Practice Test F
(64 Multiplication Facts).

Name _____

1. List the factors of 24.
(25)

2. The 25 chairs lined up in 5 equal rows. How many chairs were in each row?
(21)

Use the following information to answer problems 3 and 4:

> *There are 20 pigs that live on Ms. Anela's farm. One half of the pigs are brown, one fourth of the*
> *pigs are black, and one tenth of the pigs are pink.*

3. How many pigs are black?
(Inv. 2)

4. How many pigs are brown?
(Inv. 2)

5. Write two multiplication facts and two division facts for the fact family 5, 6, and 30.
(19)

6. What is the sum of two hundred seventy and eight hundred eighteen?
(6)

7. At the pond there are 3 teams of ducklings. If each team has 6 ducklings in it, how many ducklings are there?
(21)

8. Compare: $3 \times (2 + 5) \bigcirc 3 + (2 \times 5)$
(24)

9. Which fraction is not equal to $\frac{1}{2}$?
(23)

   A. $\frac{3}{6}$ B. $\frac{15}{30}$ C. $\frac{200}{400}$ D. $\frac{8}{17}$

10. $20 + $7.32 + $0.42
(13)

11. $7 − $3.46
(13)

12. $15 \div (10 \div 2)$
(24)

13. $8\overline{)70}$
(22)

14. $\frac{42}{7}$
(20)

15. $\begin{array}{r} \$9.04 \\ \times \quad\quad 8 \\ \hline \end{array}$
(17)

16. $7 \times 5 \times 10$
(18)

17. 728×6
(17)

18. $\begin{array}{r} 3824 \\ 276 \\ + \ 2503 \\ \hline \end{array}$
(6)

19. $\begin{array}{r} 1000 \\ - \quad 97 \\ \hline \end{array}$
(9)

20. Which number sentence below illustrates the associative property of addition?
(24)

   A. $1 + 0 = 1$ B. $1 + (3 + 5) = (1 + 3) + 5$ C. $1 + 3 = 3 + 1$

6

*Also take Facts Practice Test D
(90 Division Facts).*

Name _____

1. Eight decades is how many years?
(28)

2. Blair bought four corn dogs for $1.24 each and a soda for $0.95. How much did he spend in all?
(13)

3. What is the product of fifteen and five?
(17)

4. Britt, Sven, Marsha, Blaine, and Nina equally shared twenty frozen treats. How many frozen treats did each person have?
(21)

5. What is the largest three-digit even number that uses the digits 5, 4, and 7?
(2)

6. Think of an odd number. Multiply it by 3. Is the product odd or even?
(2)

7. If Amber is seventh in line, how many people are in front of her?
(7)

8. What temperature is shown on this thermometer?
(27)

9. List the factors of 27.
(25)

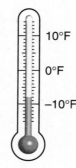

10°F

0°F

−10°F

10.
(9)
$$\begin{array}{r} 3738 \\ - 1876 \\ \hline \end{array}$$

11.
(13)
$$\begin{array}{r} \$40.00 \\ - \$\ 4.87 \\ \hline \end{array}$$

12.
(6)
$$\begin{array}{r} 23 \\ 72 \\ 84 \\ 15 \\ +\ \ 2 \\ \hline \end{array}$$

13. $237 \div 8$
(26)

14. $\dfrac{400}{5}$
(26)

15. $6\overline{)\$9.24}$
(26)

16. 60×35
(29)

17. $4 \times (5 + 9)$
(24)

18.
(17)
$$\begin{array}{r} \$8.63 \\ \times\ \ \ \ \ 6 \\ \hline \end{array}$$

19. Compare: $\dfrac{1}{2} \bigcirc \dfrac{7}{16}$
(23)

20. If it is afternoon, what time is shown by this clock?
(28)

7

Also take Facts Practice Test E
(90 Division Facts).

Name _____

1. Draw a pair of intersecting lines that are perpendicular.
(31)

2. If it is morning, what time is shown by this clock?
(28)

3. How many years were there from 1966 to 1987?
(35)

4. List the factors of 36.
(25)

5. The fish tank in Mashika's room holds 12 gallons of water. Four quarts equals one gallon. How many
(21) quarts of water does Mashika's fish tank hold?

6. Round 34 to the nearest ten.
(33)

7. Round 866 to the nearest hundred.
(33)

8. Compare: 30 − (15 − 5) ◯ (30 − 15) − 5
(24)

9. What fraction of this square is shaded?
(30)

10. What percent of this square is shaded?
(30)

11. Which of these angles is an acute angle?
(32)

A.　　　　　　　　　　　B.　　　　　　　　　　C.

12. The arrow is pointing to what number on this number line?
(27)

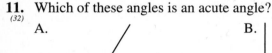

30　　　　　　50

13. $7.26 + $14 + $0.35
(13)

14. $10 − $3.26
(13)

15. 7 × 15 × 20
(18)

16. $3.75 × 30
(29)

17. 7)‾1‾4‾3‾5‾
(34)

18. 427 ÷ 6
(26)

19. $\frac{2000}{4}$
(34)

20. Which figure is not a polygon?
(32)

A.　　　　　　　B.　　　　　　　C.　　　　　　　D.

8

*Also take Facts Practice Test D
(90 Division Facts).*

Name _____

1. In Mary's garden there are 9 more red roses than yellow roses. If there are 14 yellow roses, how many
(35) red roses are there?

2. To what mixed number is the arrow pointing?
(38)

3. Which of these figures is a quadrilateral?
(32)

A.   B.   C.   D.

4. Draw an obtuse triangle.
(36)

5. Compare these fractions. Draw and shade two congruent circles to show the comparison.
(39)

$$\frac{1}{2} \bigcirc \frac{1}{3}$$

6. Round 329 to the nearest hundred.
(33)

7. What temperature is shown on this thermometer?
(27)

8. When Suzie finished page 112 of a 238-page book,
(11) she still had how many pages to read?

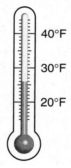

9. What fraction of this group of triangles is shaded?
(30)

10. $679 + $0.43 + $3.80
(13)

11. 7732 − 2857
(9)

12. 249 × 400
(29)

13. 7 × 15 × 3
(18)

14. 6)$6.18
(34)

15. $\frac{161}{7}$
(26)

16. 481 ÷ 3
(34)

17. $100 − $27.32
(13)

18. 200 − (20 − 2)
(24)

19. Which two triangles below are congruent?
(32)

A.   B.   C.   D.

20. According to this calendar, what is the date of the
(28) third Thursday in May 2024?

May 2024						
S	M	T	W	T	F	S
			1	2	3	4
5	6	7	8	9	10	11
12	13	14	15	16	17	18
19	20	21	22	23	24	25
26	27	28	29	30	31	

TEST

9

Also take Facts Practice Test F (64 Multiplication Facts).

Name _____

1. Which of these angles could measure 90°?
(Inv. 4)

A.    B.   C.

2. Use an inch ruler to measure the length of this segment to the nearest eighth of an inch:
(44)
_____

3. A new hat costs $15.28. Wade has $11.29. How much more money does he need to buy the hat?
(11)

4. What mixed number is half of 43?
(43)

5. Which of the quadrilaterals below is not a parallelogram?
(45)

A.   B.   C.   D.

6. What mixed number names the number of shaded pentagons?
(40)

7. A key that is 70 millimeters long is how many centimeters long?
(44)

8. One quarter plus three nickels make up what percent of a dollar?
(30)

9. Compare: $\frac{1}{4} \bigcirc \frac{1}{10}$
(39)

10. 3012 − 2877
(9)

11. $8.04 × 30
(29)

12. 7 × 5 × 7
(18)

13. 4)$51.40
(26)

14. 3483 ÷ 9
(34)

15. Use short division: 7)5250
(42)

16. $1\frac{1}{5} + 4\frac{4}{5}$
(41)

17. $\frac{8}{6} - \frac{3}{6}$
(41)

18. $\frac{1}{5} - \frac{1}{5}$
(41)

19. $40 − ($8.79 + $17 + $0.83)
(13, 24)

20. Jojo has 4 bags of beans. If each bag has 30 beans in it, how many beans does Jojo have?
(21)

Saxon Math 6/5—Homeschool

231

10

Also take Facts Practice Test F
(64 Multiplication Facts).

Name _____

1. Fred is 5 years older than Tomaz. Tomaz is 3 years older than Jean. Fred is 37 years old. How old is Jean?
(49)

2. Write the standard form for $(8 \times 1000) + (7 \times 100) + (9 \times 10)$.
(48)

3. Duc is 5 feet 4 inches tall. How many inches tall is Duc?
(47)

4. There are 14 children in one line and 22 children in the other line. If the two lines were made even, how many children would be in each line?
(50)

5. Draw a rectangle and shade three fourths of it. What percent of the rectangle is shaded?
(30, 37)

6. Use a ruler to find the length of this line segment to the nearest quarter inch:
(44)

_____

7. Seven centimeters equals how many millimeters?
(44)

8. Use digits to write the number four hundred seventy-two thousand, three hundred thirty-eight.
(7)

9. $6.59 + $14.32 + $32
(13)

10. $100 − $77.35
(13)

11. 327
(17) × 6

12. $9.02
(29) × 30

13. 4020
(9) − 327

14. 725 ÷ 6
(34)

15. 7)3000
(26)

16. $\dfrac{416}{4}$
(34)

17. $\dfrac{5}{7} - \dfrac{5}{7}$
(41)

18. $4\dfrac{2}{5} + 1\dfrac{1}{5}$
(41)

19. $3\dfrac{2}{3} - 2\dfrac{1}{3}$
(41)

20. Use the information given in this bar graph to answer the question below.
(Inv. 5)

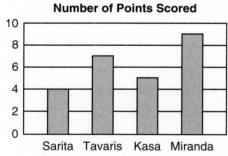

Number of Points Scored

Miranda scored how many more points than Sarita?

11

Also take Facts Practice Test G
(48 Uneven Divisions).

Name _____

1. Chinue has four jars. The jars contain 3 bugs, 8 bugs, 7 bugs, and 6 bugs respectively. If Chinue were to
(50) rearrange the bugs so that each jar had the same number, how many bugs would be in each jar?

2. How many is $\frac{3}{4}$ of two dozen?
(46)

3. How many years were there from 1066 to 1215?
(35)

4. Round 923 to the nearest ten.
(33)

5. The circle has a radius of 20 mm. What is the diameter of the circle in centimeters?
(44, 53)

6. Find the missing number: $15 \times 1 = 15 + y$
(10)

7. What is the perimeter of this rectangle?
(53)

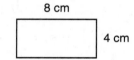

8 cm

4 cm

8. How many degrees does the minute hand of a clock turn in 15 minutes?
(Inv. 4)

9. Write the standard form for $(7 \times 100{,}000) + (2 \times 10{,}000) + (9 \times 100) + (1 \times 1)$.
(48)

10. Which triangle appears to be equilateral?
(36)

A. B. C. D.

11. $2714 + 3217 + 78$
(6)

12. $7623 - 2849$
(9)

13. $\begin{array}{r} \$2.76 \\ \times \quad 30 \end{array}$
(29)

14. $\begin{array}{r} 53 \\ \times\ 24 \end{array}$
(51)

15. $\frac{4256}{7}$
(34)

16. $20\overline{)660}$
(54)

17. $6 + \frac{1}{4}$
(43)

18. $5\frac{1}{2} - 2$
(43)

19. Draw a circle. Shade all but $\frac{3}{4}$ of it. What percent of the circle is shaded?
(30, 37)

20. Use words to name 32876534.
(52)

12

Also take Facts Practice Test F
(64 Multiplication Facts).

Name _____

1. The spinner on the right is divided into five equal-sized sectors. What is
(57) the probability that the spinner will stop on a number greater than two?

2. The divisor is 6. The dividend is 354. What is the quotient?
(20)

3. Use digits to write the number fifty-two million, six hundred
(52) eighty-seven thousand, three hundred twenty.

4. Which digit is in the hundred millions place in 843,576,324?
(52)

5. In the four boxes there are 17 pens, 7 pens, 11 pens, and 5 pens. If the pens were rearranged so that each
(50) box had the same number of pens, how many pens would be in each box?

6. What is the perimeter of this triangle?
(53)

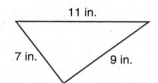

7. The radius of this circle is 8 cm. What is its diameter?
(53)

8 cm

8. Two fifths of the kittens are females. What fraction of the kittens are males?
(60)

9. The weather forecast states that the chance of rain is 75%. This means that rain is
(57)
   A. impossible   B. unlikely   C. likely   D. certain

10. Jasper missed $\frac{1}{5}$ of the 15 free throws. How many free throws did he miss?
(46)

11.   7832
(6)
       273
  + 917

12.   2420
(9) − 1903

13.   239
(55) × 621

14.   805
(56) × 220

15. 763 ÷ 10
(54)

16. Divide and write the quotient as a mixed number: $7\overline{)7023}$
(58)

17. $30\overline{)\$3.90}$
(54)

18. $7\frac{3}{5} + 2\frac{2}{5}$
(59)

19. $1 - \frac{1}{5}$
(59)

20. $30 − ($17 + $9.78 + $0.23)$
(13, 24)

13

Also take Facts Practice Test G
(48 Uneven Divisions).

Name _____

1.
(50) In one basket there are 17 oranges. In the other basket there are 23 oranges. If the oranges are rearranged so that there are the same number of oranges in each basket, then how many oranges will be in each basket?

2.
(60) Camden has eaten $\frac{1}{5}$ of his peach pie. What fraction of his pie is left to eat?

3.
(28, 35) From 1826 to 1956 was how many decades?

4.
(62) Estimate the product of 66 and 22 by rounding both numbers to the nearest ten before multiplying.

5.
(33) Round 1239 to the nearest hundred.

6.
(32, 61) Which angle appears to be an acute angle?

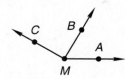

A. $\angle AMB$ B. $\angle AMC$ C. $\angle BMC$

7.
(38) On this number line, the arrow is pointing to what mixed number?

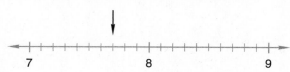

8.
(Inv. 6) If a number cube is rolled once, what is the probability of rolling a five?

9.
(53) (a) What is the length of this rectangle?

(b) What is the perimeter of this rectangle?

5 cm

3 cm

10. $9m = 63$
(18)

11. $\frac{3}{5} + \frac{2}{5}$
(59)

12. $7 - 1\frac{2}{3}$
(63)

13. $3 - \frac{1}{5}$
(63)

14. $3821 - 930$
(9)

15. $\$20 - (\$8 + \$5.47)$
(13, 24)

16. $\begin{array}{r} 125 \\ \times\ 432 \\ \hline \end{array}$
(55)

17. $5633 \div 8$
(34)

18. $20\overline{)340}$
(54)

19.
(64) What is the place value of the 8 in $\$19.84$?

20.
(58) Divide and write the quotient as a mixed number: $\frac{59}{8}$

14

Also take Facts Practice Test C
(100 Multiplication Facts).

Name _____

1. Addar bought three muffins for $0.74 each. If he gave the clerk $5.00, how much money should he get back?
₍₄₉₎

2. List the factors of 16 that are also factors of 24.
₍₂₅₎

3. Name the shaded part of this square
₍₆₇₎
   (a) as a fraction
   (b) as a decimal number

4. Use words to write the decimal number 76.25.
₍₆₈₎

5. Write the fraction $\dfrac{87}{100}$ as a decimal number.
₍₆₇₎

6. Use digits to write the decimal number fourteen and twenty-three hundredths.
₍₆₈₎

7. Which digit in 247.38 is in the tenths place?
₍₆₄₎

8. Find the length of the line segment below to the nearest tenth of a centimeter:
₍₆₆₎

9. Compare: 0.4 $\bigcirc$ 0.04
₍₆₉₎

10. Estimate the product of 27 and 41.
₍₆₂₎

11. 274 + 38 + 3519 + 6 + 12
₍₆₎

12.   30,000
₍₉₎ − 21,325

13.   227
₍₅₆₎ × 160

14. $\dfrac{4248}{6}$
₍₃₄₎

15. 20 × 35 × 3
_(18, 29)

16. 5$\overline{)850}$
₍₃₄₎

17. $95.60 ÷ 40
₍₅₄₎

18. $7 - \left(5\dfrac{2}{3} - 4\right)$
_(24, 63)

19. $1\dfrac{1}{5} + 2\dfrac{2}{5} + 3\dfrac{2}{5}$
₍₅₉₎

20. Divide and write the quotient as a mixed number: $\dfrac{29}{6}$
₍₅₈₎

T E S T

15

Also take Facts Practice Test H
(60 Improper Fractions to Simplify).

Name _____

1. Arrange these fractions in order from least to greatest:
(23, 59)

$$\frac{1}{4}, \frac{6}{6}, \frac{4}{8}, \frac{2}{3}$$

2. Write the fraction $\frac{8}{3}$ as a mixed number.
(75)

3. Twelve inches equals one foot. Three feet equals one yard. Eight yards equals how many inches?
(74)

4. Use digits to write the decimal number thirteen and fifty-two hundredths.
(68)

5. Which digit in 456.29 is in the hundredths place?
(64)

6. Name the shaded part of this square
(71)
   (a) as a fraction
   (b) as a decimal number
   (c) as a percent

7. Round 8734 to the nearest hundred.
(33)

8. Compare: 8.42 ◯ 8.420
(70)

9. Divide 873 by 40 and write the quotient as a mixed number.
(58)

10. $9.28 + 97¢ + $2
(70)

11. $1.22 − 74¢
(70)

12.   8.26
(73)   1.4
   + 0.72

13.   4.892
(73) − 1.9

14. $0.46
(17) ×   3

15. $8 - \left(7\frac{4}{5} - \frac{3}{5}\right)$
(24, 63)

16. $2\frac{3}{5} + 3\frac{2}{5}$
(59)

17. The length of $\overline{PQ}$ is 50 mm. Segment QR is 20 mm long. Find the length of $\overline{PR}$ in **centimeters**.
(61, 74)

18. Draw a parallelogram with all four sides equal in length.
(45)

A rectangular room is 6 yards long and 5 yards wide. Use this information to answer problems 19 and 20.

19. What is the perimeter of the room?
(53)

20. What is the area of the room?
(72)

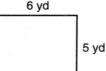

6 yd

5 yd

16

Also take Facts Practice Test H
(60 Improper Fractions to Simplify).

Name _____

1. Rudo bought three model airplanes for $3.40 each and a jar of green paint for 85¢. What was the total cost
(49) of the items?

2. Write 3^4 as a whole number.
(78)

3. Write $\dfrac{9}{100}$ as a decimal number.
(67)

4. How many centimeters are in 7.2 kilometers?
(74)

5. One ton equals 2000 pounds. A six ton elephant weighs how many pounds?
(77)

6. Which digit in 287.56 is in the same place as the 4 in 38.4?
(64)

7. Compare: $\dfrac{1}{2} \times \dfrac{6}{6} \bigcirc \dfrac{2}{4}$
(79)

8. Write a fraction equal to $\dfrac{3}{5}$ that has a denominator of 10.
(79)

9. The first prime number is 2. What are the next four prime numbers?
(80)

10. Find the length of this segment to the nearest tenth of a centimeter:
(66)

11. $3 + $2.87 + 84¢ + $16 + 6¢
(70)

12. 21.72
(73)   2.2
      + 17.3

13.   87.56
(73)   −   8.7

14. $\dfrac{9510}{30}$
(54)

15. $6.39
(17)   ×   4

16. 20 × 40 × 60
(18, 29)

17. 4)$57.00
(26)

18. $\dfrac{4}{5} + \dfrac{2}{5}$
(75)

19. $\dfrac{2}{3} \times \dfrac{1}{3}$
(76)

20. *EF* is 34 mm. *EG* is 95 mm. Find *FG*.
(61)

Saxon Math 6/5—Homeschool

17

Also take Facts Practice Test H
(60 Improper Fractions to Simplify).

Name _____

1. Each of the following numbers divides 410 without a remainder except
(22, 42)
    A. 2   B. 3   C. 5   D. 10

Refer to quadrilateral *ABCD* to answer problems 2 and 3.

2. Angle *BAD* is acute. Which angle is obtuse?
(32, 61)

3. Which segment is perpendicular to $\overline{CD}$?
(31, 61)

4. What is the name of this shape?
(83)
    A. cube   B. cone

    C. pyramid   D. cylinder

5. (a) What number is $\frac{1}{3}$ of 24?   (b) What number is $\frac{2}{3}$ of 24?
(46)

6. One sixth of the 18 houses were brick. How many houses were brick?
(46)

7. (a) Find the greatest common factor (GCF) of 16 and 24.
(82)

    (b) Use the GCF of 16 and 24 to reduce $\frac{16}{24}$.

8. How many quarts are in three gallons?
(85)

9. Reduce each fraction:
(81)

    (a) $\frac{4}{16}$ (b) $\frac{14}{21}$ (c) $\frac{3}{6}$

10.   37.85
(73)   6.629
     + 263.4

11.   29.684
(73)   − 17.79

12. 6^3
(78)

13. $4\overline{)\$35.00}$
(26)

14.   7010
(9)   − 3976

15.   450
(56) × 803

16. $\frac{5220}{60}$
(54)

17. $5\frac{4}{5} + 2\frac{4}{5}$
(75)

18. $5 - \left(\frac{1}{4} + 2\right)$
(24, 63)

19. Compare: $\frac{2}{5} \times \frac{4}{4} \bigcirc \frac{2}{5} \times \frac{2}{2}$
(79)

20. What is the area of this square?
(66, 72)

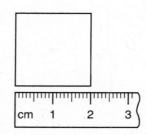

*Also take Facts Practice Test I
(40 Fractions to Reduce).*

Name _____

1. The movie starts at 7:45 p.m. It takes Jermael 10 minutes to walk to the theater. At what time should he
(28, 49) start for the movie if he wants to get there 15 minutes early?

2. Write the fraction $\dfrac{12}{5}$ as a mixed number.
(75)

3. Reduce: $\dfrac{12}{15}$
(90)

4. A quart of water is how many ounces of water?
(85)

Refer to quadrilateral *ABCD* to answer problems 5 and 6.

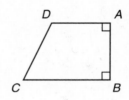

5. Which segment is parallel to $\overline{AD}$?
(31, 61)

6. Which angle is an acute angle?
(32, 61)

7. A soup can has the shape of a
(83)
    A. pyramid   B. cylinder   C. sphere   D. rectangular solid

8. What number is $\dfrac{4}{5}$ of 30?
(86)

9. What is the median of these seven test scores?
(84)
$$95, 80, 100, 80, 90, 95, 80$$

10. Write a fraction equal to $\dfrac{3}{4}$ that has a denominator of 12.
(79)

11. Write 0.625 with words.   **12.** $84{,}016 - 8127$ **13.** 907×120
(68)   (9)   (56)

14. $4\overline{)\$15.24}$ **15.** $\dfrac{2}{5} \times \dfrac{1}{3}$ **16.** $\sqrt{49}$
(26)   (76)   (89)

17. $\dfrac{3}{5} \div \dfrac{3}{5}$ **18.** $\dfrac{7}{8} - \dfrac{3}{8}$ **19.** $2\dfrac{1}{6} + 3\dfrac{1}{6}$
(87)   (90)   (90)

20. What is the perimeter of this square?
(53, 66)

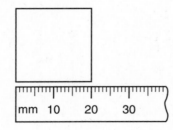

19

Also take Facts Practice Test I
(40 Fractions to Reduce).

Name _____

1. What is the reciprocal of $\frac{5}{4}$?
(95)

2. Write fractions equal to $\frac{1}{3}$ and $\frac{1}{5}$ with denominators of 15. Then add the fractions.
(79)

3. Makala had 2 dozen cookies, which she laid in 6 rows. How many cookies were in each row?
(21)

4. (a) What number is $\frac{1}{5}$ of 70?
(86)

(b) What number is $\frac{3}{5}$ of 70?

5. Name the shaded part of the circle
(71)
(a) as a reduced fraction

(b) as a decimal number

(c) as a percent

6. List these numbers in order from least to greatest:
(69)

$$0.3, \frac{2}{3}, 0$$

7. Write the length of this segment
(66)
(a) as a number of centimeters

(b) as a number of millimeters

8. Write $\frac{24}{14}$ as a mixed number with the fraction reduced.
(91)

9. $\frac{2}{3} = \frac{x}{12}$
(79)

10.
(73)
$$\begin{array}{r} 27.2 \\ 9.43 \\ + 17.5 \\ \hline \end{array}$$

11.
(9)
$$\begin{array}{r} 5201 \\ - 3839 \\ \hline \end{array}$$

12.
(51)
$$\begin{array}{r} \$6.25 \\ \times \quad 24 \\ \hline \end{array}$$

13. $10\overline{)\$22.00}$
(54)

14. $\sqrt{36}$
(89)

15. $16\overline{)832}$
(94)

16. $7\frac{7}{8} - 2\frac{5}{8}$
(90)

17. $2\frac{4}{5} + 3\frac{4}{5}$
(91)

18. $\frac{3}{4} \times 5$
(86)

19. $\$7.83 + 59¢ + \$14$
(70)

20. $14 \times 35¢$
(70)

Also take Facts Practice Test J
(50 Fractions to Simplify).

Name _____

Use the following information to answer problems 1 and 2.

Sue has a number-changing machine that works in such a way that when she puts in a 24, an 8 comes out. When she puts in a 12, a 4 comes out, and when she puts in a 3, a 1 comes out.

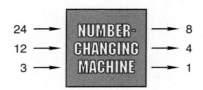

24 → NUMBER- → 8
12 → CHANGING → 4
3 → MACHINE → 1

1. What rule does Sue's machine use?
(Inv. 7)
    A. It subtracts 7.   B. It divides by 4.   C. It multiplies by 2.   D. It divides by 3.

2. If Sue puts in a 9, what number will come out?
(Inv. 7)

3. There are 12 girls and 15 boys in the painting. What is the ratio of girls to boys in the painting?
(97)

4. The snail crawled 12 centimeters. Twelve centimeters is how many millimeters?
(74)

5. *WX* is 3.4 cm. *XY* is 1.4 cm. *YZ* equals *XY*. Find *WZ*.
(61, 73)

      W   X   Y   Z

6. (a) An ounce is what fraction of a pint?
(74, 76)

    (b) A pint is what fraction of a quart?

    (c) Use the answers from (a) and (b) above to determine what fraction of a quart an ounce represents.

7. Estimate the product of 325 and 671.
(62)

8. Simplify this decimal number: 00021.090
(100)

9. What is the reciprocal of $\frac{9}{4}$?
(95)

10. Reduce: $8\frac{12}{21}$
(91)

11. $8.5 + 0.31 + 2$
(99)

12. 2^6
(78)

13. $7.982 - 0.35$
(73)

14. $10 \times 97¢$
(70)

15. $\sqrt{64}$
(89)

16. $34\overline{)700}$
(92)

17. $9 - \left(3\frac{2}{5} + 1\frac{1}{5}\right)$
(24, 63)

18. $\frac{3}{5} \div \frac{2}{3}$
(96)

19. $\frac{5}{6} + \frac{5}{6}$
(91)

20. $\frac{2}{3} \times 2$
(86)

21

Also take Facts Practice Test J
(50 Fractions to Simplify).

Name _____

1. Round $12.64 to the nearest dollar.
(104)

2. (a) Round 18.3 to the nearest whole number.
(101, 104)

(b) Round $7\frac{2}{3}$ to the nearest whole number.

3. Arrange these numbers in order from least to greatest:
(69)

$$0.8, 0.2, 0$$

4. Two thirds of the 24 books are fiction. How many books were are fiction?
(46)

5. The length of $\overline{AD}$ is 7.9 cm. The length of $\overline{AB}$ is 2 cm. The length of $\overline{BC}$ is 3.8 cm. Find the length of $\overline{CD}$.
(61)

6. A basketball has the shape of what geometric solid?
(83)

7. Which point is located at $(-3, 2)$?
(Inv. 10)

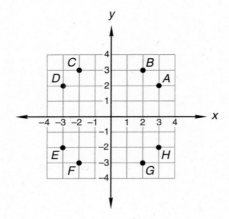

8. $9.75 + 2.3 + 4$
(99)

9. $5.17 - 0.6$
(73)

10. $5 - 3.4$
(102)

11. 144×24
(51)

12. $5 - \left(2\frac{7}{8} - \frac{5}{8}\right)$
(63, 90)

13. $2\frac{6}{8} + 1\frac{6}{8}$
(91)

14. $\dfrac{7014}{7}$
(34)

15. $426 \div 30$
(54)

16. $\dfrac{1}{4} \div \dfrac{1}{3}$
(96)

17. $\dfrac{3}{4} \times 4$
(86, 91)

18. Find the volume of this rectangular solid.
(103)

19. Draw a square and show its lines of symmetry.
(105)

20. The denominator of $\frac{9}{10}$ is 10. Write a fraction equal to $\frac{1}{2}$ that also has a denominator of 10 and subtract that fraction from $\frac{9}{10}$. Then reduce the answer.
(79, 90)

22

Also take Facts Practice Test K
(30 Percents to Write as Fractions).

Name _____

1. Which digit in 2.7182 is in the thousandths place?
(106)

2. The swimming pool is 25 yards long. How many feet long is the swimming pool?
(74)

3. Which of these figures has no line of symmetry?
(105)

A.
B.
C.
D.

4. There are 8 girls and 16 boys in the play. What is the ratio of girls to boys in the play?
(97)

5. Write 10% as a reduced fraction.
(71, 90)

6. Sixteen of the 50 cars on the lot were built before 1995. What percent of the cars on the lot were built before 1995?
(107)

7. Round to the nearest whole number:
(101, 104)

(a) $7\dfrac{4}{9}$

(b) 32.701

8. Compare: $0.036 \bigcirc 0.36$
(106)

9. What is the perimeter of this square?
(53)

10. What is the area of this square?
(72)

13 in.

11. $14 + 6.82 + 0.927 + 3$
(99)

12. $4.6 - 1$
(99)

13. $7 - 2.91$
(102)

14. 2.4×0.3
(109)

15.
(110)
$$\begin{array}{r} 0.16 \\ \times\ \ 0.6 \\ \hline \end{array}$$

16. $2\dfrac{2}{5} + \left(7 - 3\dfrac{2}{5}\right)$
(63, 90)

17. $\dfrac{4}{5} \times \left(5 \times \dfrac{1}{4}\right)$
(86, 91)

18. $2\dfrac{2}{3} + 2\dfrac{2}{3}$
(91)

19. $\dfrac{9}{10} \div 3$
(90, 96)

20. What is the volume of a box of cereal with the dimensions shown?
(103)

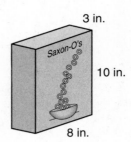

3 in.

Saxon-O's

10 in.

8 in.

Also take Facts Practice Test K
(30 Percents to Write as Fractions).

Name _____

1. Estimate the sum of 6.38, 5.83, and 7.66 by rounding each number to the nearest whole number
(104) before adding.

2. Which of these is **not** equal to $\frac{1}{2}$?
(71)

   A. 0.50   B. $\frac{7}{14}$ C. 40%   D. $\frac{5}{10}$

3. What is the temperature shown on this
(98) thermometer?

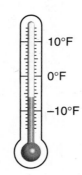

Use the following graph to answer problems 4 and 5.

4. What letter names the point at $(-1, -2)$?
(Inv. 10)

5. Write the coordinates of point *G*.
(Inv. 10)

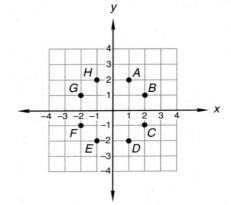

6. What is the mean of 13, 10, 9, 5, and 8?
(84)

7. The spinner at right is divided into 6 congruent sectors. What is the probability
(57) that the spinner will stop on an odd number?

8. Refer to the hexagon at the right to
(53, 115) answer the following questions. All
angles are right angles.
(a) What is the perimeter of this hexagon?

(b) What is the area of this hexagon?

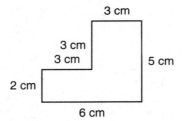

9. Write $2\frac{3}{4}$ as an improper fraction.
(113)

10. Find the least common multiple of 8 and 10.
(112)

11. $5.82 + 17 + 0.186 + 13$
(99)

12. $10 - (8.69 - 4)$
(102)

13. 4.7×12
(109)

14. $18\overline{)534}$
(92)

15. 0.5432×100
(111)

16. 0.2×0.01
(110)

17. $2\frac{7}{8} + 2\frac{7}{8}$
(91)

18. $5\frac{5}{8} - 1\frac{3}{8}$
(90)

19. $\frac{3}{4} \times \frac{2}{3}$
(76, 90)

20. $\frac{5}{6} \div \frac{2}{3}$
(96)

Recording Forms

The five optional recording forms in this section may be photocopied to provide the quantities needed by you and your student.

Recording Form A: Facts Practice
This form helps your student track his or her performances on Facts Practice Tests throughout the year.

Recording Form B: Lesson Worksheet
This single-sided form is designed to be used with daily lessons. It contains a checklist of the daily lesson routine as well as answer blanks for the Warm-Up and Lesson Practice.

Recording Form C: Mixed Practice Solutions
This double-sided form provides a framework for your student to show his or her work on the Mixed Practices. It has a grid background and partitions for recording the solutions to thirty problems.

Recording Form D: Scorecard
This form is designed to help you and your student track scores on daily assignments and cumulative tests.

Recording Form E: Test Solutions
This double-sided form provides a framework for your student to show his or her work on the tests. It has a grid background and partitions for recording the solutions to twenty problems.

A Facts Practice

Name _____

TEST	TIME AND SCORE $\diagup$ time $\diagdown$ # correct										
A 100 Addition Facts											
B 100 Subtraction Facts											
C 100 Multiplication Facts											
D **E** 90 Division Facts											
F 64 Multiplication Facts											
G 48 Uneven Divisions											
H 60 Improper Fractions to Simplify											
I 40 Fractions to Reduce											
J 50 Fractions to Simplify											
K 30 Percents to Write as Fractions											

Saxon Math 6/5—Homeschool

B   Lesson Worksheet
Show all necessary work. Please be neat.

Name _____

Date _____

Lesson _____

Warm-Up
☐ Facts Practice
☐ Mental Math
☐ Problem Solving

Review
☐ Homework Check
☐ Error Correction

Instruction
☐ Lesson
☐ Lesson Practice
☐ Mixed Practice

Facts Practice

Test:	Time:	Score:

Mental Math

a.	b.	c.	d.	e.	f.
g.	h.	i.	j.	k.	l.

Problem Solving

Strategies:
(Check any you use.)

☐ Make a chart, graph, or list.
☐ Guess and check (trial and error).
☐ Use logical reasoning.
☐ Act it out.   ☐ Draw a diagram.
☐ Make it simpler. ☐ Draw a picture.
☐ Work backward.   ☐ Find a pattern.

Lesson Practice

a.	b.	c.
d.	e.	f.
g.	h.	i.
j.	k.	l.

Saxon Math 6/5—Homeschool

C

Mixed Practice Solutions
Show all necessary work. Please be neat.

Name _____

Date _____

Lesson _____

2.	3.
5.	6.
8.	9.
11.	12.
14.	15.

16.

17.

18.

19.

20.

21.

22.

23.

24.

25.

26.

27.

28.

29.

30.

RECORDING FORM

D | Scorecard

Name _____

Date	Lesson or Test	Score	Date	Lesson or Test	Score	Date	Lesson or Test	Score	Date	Lesson or Test	Score

Saxon Math 6/5—Homeschool

E | Test Solutions

Show your work on this paper.
Do not write on the test.

Name _____

Date _____

Test _____ Score _____

2.

4.

6.

8.

10.

11.

12.

13.

14.

15.

16.

17.

18.

19.

20.

Saxon Math 6/5—Homeschool